D0221253

The Social Life of Nanotechnology

Routledge Studies in Science, Technology and Society

1 **Science and the Media**
Alternative Routes in Scientific Communication
Massimiano Bucchi

2 **Animals, Disease and Human Society**
Human-Animal Relations and the Rise of Veterinary Medicine
Joanna Swabe

3 **Transnational Environmental Policy**
The Ozone Layer
Reiner Grundmann

4 **Biology and Political Science**
Robert H Blank and Samuel M. Hines, Jr.

5 **Technoculture and Critical Theory**
In the Service of the Machine?
Simon Cooper

6 **Biomedicine as Culture**
Instrumental Practices, Technoscientific Knowledge, and New Modes of Life
Edited by Regula Valérie Burri and Joseph Dumit

7 **Journalism, Science and Society**
Science Communication between News and Public Relations
Edited by Martin W. Bauer and Massimiano Bucchi

8 **Science Images and Popular Images of Science**
Edited by Bernd Hüppauf and Peter Weingart

9 **Wind Power and Power Politics**
International Perspectives
Edited by Peter A. Strachan, David Lal and David Toke

10 **Global Public Health Vigilance**
Creating a World on Alert
Lorna Weir and Eric Mykhalovskiy

11 **Rethinking Disability**
Bodies, Senses, and Things
Michael Schillmeier

12 **Biometrics**
Bodies, Technologies, Biopolitics
Joseph Pugliese

13 **Wired and Mobilizing**
Social Movements, New Technology, and Electoral Politics
Victoria Carty

14 **The Politics of Bioethics**
Alan Petersen

15 **The Culture of Science**
How the Public Relates to Science Across the Globe
Edited by Martin W. Bauer, Rajesh Shukla and Nick Allum

16 **Internet and Surveillance**
The Challenges of Web 2.0 and Social Media
Edited by Christian Fuchs, Kees Boersma, Anders Albrechtslund and Marisol Sandoval

17 **The Good Life in a Technological Age**
Edited by Philip Brey, Adam Briggle and Edward Spence

18 **The Social Life of Nanotechnology**
Edited by Barbara Herr Harthorn and John W. Mohr

The Social Life of Nanotechnology

Edited by Barbara Herr Harthorn
and John W. Mohr

NEW YORK LONDON

First published 2012
by Routledge
711 Third Avenue, New York, NY 10017

Simultaneously published in the UK
by Routledge
2 Park Square, Milton Park, Abingdon, Oxon OX14 4RN

Routledge is an imprint of the Taylor & Francis Group, an informa business

Library of Congress Cataloging-in-Publication Data

The social life of nanotechnology / edited by Barbara Herr Harthorn and John W. Mohr.
p. cm. — (Routledge studies in science, technology and society ; 18)
Includes bibliographical references and index.
1. Nanotechnology—Social aspects. I. Harthorn, Barbara Herr, 1951– II. Mohr, John.
T14.5.S63826 2012
303.48'3—dc23
2012002812

ISBN: 978-0-415-89905-5 (hbk)
ISBN: 978-0-203-10647-1 (ebk)

Typeset in Sabon
by IBT Global.

Printed and bound in the United States of America on sustainably sourced paper by IBT Global.

Contents

Figures

Foreword

John Seely Brown

I was delighted to be asked to write the foreword to *The Social Life of Nanotechnology*, an encompassing collection of scholarly works touching nearly every aspect of the social currents underlying the launching of this field, its radically cross-disciplinary nature, and the crucial issue of how to engage the public in a meaningful dialogue about the risks and opportunities that this promising field might produce. Speaking personally, nearly 12 years ago I had become embroiled in the frenzy around the first major dystopian article about nanotechnology, Bill Joy's (in)famous article in *Wired*, "Why the Future Doesn't Need Us." Joy postulated that self-replicating nanobots could run amuck and consume us all. Because he was no Luddite but rather an awe-inspiring technologist, the *digiratti* became engrossed in debating the issue along with the popular press.

In response to this near hysteria, Paul Duguid and I wrote a short article, extending the thesis of our book *The Social Life of Information*, called "Don't Count Society Out" (Brown & Duguid, 2001). Our purpose was to call attention to the 6-D thinking (*demassification*, *decentralization*, *disintermediation*, *despacialization*, *disaggregation*, and *demarketization*) found in most futurology and how it creates a kind of tunnel vision that trivializes the co-evolutionary forces at play between society and technology. I had expected the appearance of this article would lead to the usual calls from journalists, but what I had not expected was that when I said I wanted to give a balanced account of the promises and risks of this emerging field, one journalist, whom I knew from prior interviews, said she completely understood my intent but that the final copy would be out of her hands and that the editor would unquestionably cut all comments about its promises and focus solely on its risks. I thanked her for being so honest and declined the interview. Nevertheless this episode brought home to me the growing challenges of a world where technological disruption is happening more rapidly than ever and society's understanding of technology is falling farther and farther behind. Our current institutions, such as the press, are not inventing ways to help bridge the gap between what we know and what we need to know. That is why I welcomed this new book so much.

Nanotechnology is a scientific and technological endeavor that is still emerging, one that is creating new materials with nearly magical properties and new systems that re-engineer biological mechanisms to perform extraordinary feats. Indeed, this field stretches our imagination in provocative ways and as such may serve as a path to reignite a passion for learning around STEAM (science, technology, arts, and mathematics) in today's students—much like what happened with NASA's mission to the moon. But unlike that period with its fixed and audacious goal, this one has no one specific aim. Each community or country can imagine its own end point. For example, just imagine the idea of building a space elevator out of carbon nanotube ribbons to a geostationary orbit. Might this be doable, given the properties of carbon nanotube fibers? Curiously, what has started to emerge from parts of the nanotechnology community is a new form of theoretical engineering that explores what *in principle* might be possible given our understanding of quantum mechanics, and so on. Indeed, in my own research center (PARC), Ralph Merkel with his colleague Eric Drexler were deeply engaged in determining if we could build a CAD/CAM system for assembling a set of fundamental building blocks (e.g., nano-gears) for building a universal assembler inspired by the classical computer science ideas of Von Neumann. These endeavors had to confront the theoretical boundaries of a practice, not just the boundaries of a technology, turning on such issues as the inherent thermal noise in any such system.

Challenging and mesmerizing as the technological feats of this field may be, they may not be as challenging as are the issues of finding ways to scaffold the public's understanding of what is actually being produced, what the risks are, and what might need to be regulated. Said differently, how best can the co-evolution of the social with the technological be expedited? In the nanotechnology realm this challenge is exacerbated by the speed of technological advancement, the newness of the mechanisms involved in constructing nano artifacts, and the novelty of the end products themselves. This is where several of the key chapters of this book take aim, detailing systematic ways to engage the public in dialogue around the pros and cons of particular nanotechnologies. From my own experience I would suggest that as daunting as these challenges are, we are also beginning to invent new ways to use social technologies to enhance public engagement around key societal issues. Two examples warrant special attention. The first stems from the use of an Internet annotation technology, Reframe It (2012), that facilitates groups of people annotating web documents and that, in turn, has been used to amplify and scale various methods for conducting public deliberation. The technology by itself is not so surprising, but when coupled to carefully designed social protocols for deliberation, it shows good promise. A second example, currently in use by the U.S. Patent Office, called Peer-to-Patent (see Noveck, 2010), has created a set of social protocols for using social media to collect judgments on the prior art of the invention covered by the patent application. As in the spirit of this book, the digital tools themselves are only a tiny part of the solution. It is the close coupling of the social network technology along with well-designed social protocols that are showing so much promise in enabling meaningful public discourse.

REFERENCES

Brown, J. S., & Duguid, P. (2001). Don't count society out. In P. J. Denning (Ed.), *The invisible future: The seamless integration of technology into everyday life*, pp. 117–144. New York: McGraw-Hill.

Noveck, E. (2010). *Wiki government: How technology can make government better, democracy stronger, and citizens more powerful.* Washington, DC: Brookings Institution Press.

Reframe It. (2012). *The deliberative corporation: Summary.* Retrieved February 6, 2012, from http://www.reframeit.com/documents/the-deliberative-corporation

Acknowledgments

The authors gratefully acknowledge the efforts of the many colleagues, friends and family who have made this volume possible. The research team leaders of the Center for Nanotechnology in Society at UC Santa Barbara (CNS-UCSB), Richard P. Appelbaum and W. Patrick McCray (along with Barbara Herr Harthorn) contributed the basic framework for the volume and helped recruit other contributors. They and CNS-UCSB fellow leaders Bruce Bimber and Christopher Newfield provided essential critical analysis of the initial framing of the volume. CNS National Advisory Board Chair, John Seely Brown, has generously lent his name, time and intellectual support to this venture, for which we thank him while yet exonerating him from any responsibility for the final product. We thank in absentia our dear colleague in the CNS and UC Center for Environmental Implications of Nanotechnology, William Freudenburg, whose untimely death in December 2010 has created an irreplaceable gap in our work and lives; yet the legacy of his important work on risk and society will endure, as is evident in many of the chapters in this volume and more particularly in the closing chapter. The authors would also like to thank Cassandra Engeman and Christine Shearer for their invaluable assistance in preparing the manuscript for submission and Kristen Nation for her help with copyediting and proofing. Finally, editor Max Novich and Senior Editorial Assistant Jennifer Morrow at Routledge have been consistently helpful and supportive editors, making our work enjoyable and productive.

1 Introduction

The Social Scientific View of Nanotechnologies

Barbara Herr Harthorn and John W. Mohr

The Social Life of Nanotechnology addresses the interconnections and tensions between technological development, the social benefits and risks of technology, and the changing political economy of a global world system as they apply to the emerging field of nanotechnologies. The basic premise, developed throughout the volume, is that nanotechnologies have an under-theorized and often invisible social life that starts with the very concept of "nanotechnology" itself which, as we show in the volume, takes on a wide range of socio-historically specific meanings around the globe, across multiple localities, institutions and collaborations, through diverse industries, research labs, and government agencies, and on into a variety of discussions within the public sphere itself. The volume looks at this process through the lenses of the social and cultural sciences, revealing a surprisingly complicated social milieu where a series of traditionally modernist scientific projects have been (and are continuously being) reassembled into new configurations that are sharply marked by their emergence within a rapidly changing, increasingly globalized, and decidedly postmodern world. As the authors in this volume explain, this results in a series of unique contradictions, tensions, and unexpected developments. We highlight three dimensions of this process in the papers collected here: the early origins of nanotechnologies, questions about the social (and political) organization of the field, and studies concerned with the cultural and subjective meanings ascribed to nanotechnologies in social settings.

We look at how nanotechnologies (like all organized social endeavors) are socially constructed or, as Latour (2005) would say, assembled from a range of different preexisting technologies, scientific theories, professional communities, streams of resources, and intellectual mandates. The assemblers are the agents of numerous communities and interests who have pushed and shoved to make this into a recognizable arena of action. Nanotechnologies emerged from a range of different post–Cold War scientific projects that were drawn together through collaborations that followed on a series of technical and cultural breakthroughs. Technological developments involved the convergence of actions and interventions at the molecular or nano scale (measured in billionths of a meter). Cultural breakthroughs arrived via scientific visions of possible futures, collective constructions of what counts as "valued research" and real-world assemblages of shared identities built from social ties, interdisciplinary scientific collaborations, and (as Mody shows in Chapter 4, this volume) time spent together at conferences. This burgeoning innovation system has taken form in tandem with the rise of the Internet, the rapid growth of modern political economic globalization, and the increased digitalization of media and meaning. In ways that we describe in the volume, this means that perhaps more than any other contemporary scientific field, nanotechnologies have been assembled through a kind of scientific and technological bricolage, an amalgamation of conjoint ideas and technologies, organized around no singular disciplinary community or intellectual site but rather by the scale at which the work is conducted and, increasingly, the surprising functionalities of new materials accessible at that scale.

Second, we focus on how the field (or fields) have been and are actively being politically conceived and socially organized. Here again, we argue that nanotechnologies, envisioned by their research and development (R&D) advocates in governments and industry as a pure and linear modernist science enterprise, have emerged fully entangled in a fractured, heterogeneous postmodern world. Their arrival coincides with a time when more traditional systems of power and authority in national, international, and professional communities are being upended and recast in new and unfamiliar ways. Thus longstanding expectations about the relationship between science, technology, the market, and the state are being actively reconfigured and reinvented, and centers of power are shifting in unexpected directions. In the United States the state has continued to play a key role in pushing for the creation and recognition of nanotechnology as a scientific arena, but this has played out just as the more traditional linkages between scientific communities, the modern research university, and capital markets are being actively renegotiated and redefined. As Johansson argues (Chapter 6, this volume) one result is that the role and status of scientific labor and indeed the very idea of traditional scientific careers are now being challenged and recast. Meanwhile, as economic fortunes have shifted, nanotechnologies are among the key tools and pathways for China, India, Brazil, Russia, and other emerging global powerhouses to generate new models of scientific development and reconfigure the way that scientific knowledge and new technologies are globally created, produced, distributed, and consumed.

Third, we focus on the social and cultural construction of nanotechnologies, asking how different publics, experts, and agents are considering, contesting, and debating the meaning and implications of these heterogeneous things—nanotechnologies—in many cases well before they emerge into the marketplace or public awareness. Issues of democratization of science include full consideration of the risks, benefits, and implications of public participation in technological decision making and are central to this discussion, and the "deliberative turn" is now a widespread global phenomenon (Whitmarsh, 2009). The work thus draws on the fundamental questions about risk, society, and modernity raised by Giddens (1990), Beck (1992a, 1992b), and Beck, Giddens, and Lash (1994), but goes beyond them as well to address contemporary issues of democratic participation. Here the questions have to do with how the fundamental goals to develop these technologies are being represented, by whom, to what aims, and how their potential consequences—opportunity costs as well as benefits, risks and burdens—are being defined (Wynne, 1992). Through what kinds of narratives do key agents frame the debate, and what types of interpretive contestations are emerging around issues of economic versus social progress, equitable distribution of potential benefits and harms, procedural justice, and many other social risks? How, in other words, does the global response to nanotechnologies reflect different patterns of hope, trust, skepticism, ambivalence, and fear? Others have considered downstream consequences of technological development (Funtowicz & Ravetz, 1992, 2003; Ravetz, 1999), but the upstream poses potentially paradigm-shifting challenges for addressing uncertainty and societal impacts of technological development (Rogers-Hayden, Mohr, & Pidgeon, 2007; Wynne, 1992). Once again, we see nanotechnology as expressing an essential contradiction of postmodern life as a variety of stakeholders maneuver to assess, define, and reshape our frames for understanding nanotechnologies, often before they are actually invented or deployed.

For these reasons we see the need for a careful and wide-ranging assessment of what we call the social life of nanotechnology. The book draws inspiration from anthropologist Appadurai's volume *The Social Life of Things* (1986), which asked how people across numerous cultural, national, and historical sites have come to attribute value and meaning to commodities and "things." To suggest that things or commodities have social lives is to ask how value is attached, how a large range of material objects in the cultural field are wanted, circulated, traded, and ascribed with social meanings through a relatively circumscribed set of social and political mechanisms. Appadurai's work and that of fellow anthropologist Kopytoff (1986) in the volume extend the social life of things far beyond their construction as commodities, arguing that social life also imbues artifacts with complex, symbolic values. Examples include the way luxury goods acquire symbolic capital far beyond their economic value (Bourdieu, 1984; Kopytoff, 1986, p. 67) and the way sacred things are taken out of commodity valuation at the end of life (Weiner, 1992). Like this volume, Appadurai's collection drew on a range of cultural scholars, social scientists, and historians. Appadurai's volume as well similarly drew focused attention to the maintenance of social power effected by and through the circulation of commodities (things).

The other main intellectual heritage to which this volume lays claim is *The Social Life of Information* (Brown & Duguid, 2000), which had as its main aim to remind scientists, technologists, business, and government that the social contexts of technologies demand close and careful attention and understanding, and that the intuitive assumptions of experts and practitioners are insufficient to that understanding. Countering similar scientific, governmental, and business intuitive (mis)understandings of social and institutional realities with carefully drawn and methodologically sophisticated empirical and theoretical research is a primary imperative of the work presented here as well.

In the volume scholars from diverse disciplines—anthropology, cultural studies, economics, environmental studies, global studies, history, political science, risk studies, social psychology, sociology, and women's studies—bring a wide range of intellectual tools to the task of analyzing the diverse social lives of nanotechnologies. By comparing and contrasting these processes in the United States with other initiatives in China and with public responses in the UK, the book also provides one of the first comparative (and global) perspectives on nanotechnologies as they emerge and take shape in a variety of forms across a range of local settings.

The first part of the book takes up the study of nanotechnology as a historically specific institutional site. Here key questions concern how nanotechnology was originally (and, in an ongoing way, continues to be) constructed as a social, political, and scientific concept, as a domain of knowledge, and as a set of technologically sculpted scientific practices. In each of these chapters, how nanotechnology is constructed (i.e., how scientific, cultural, technological, and economic resources are organized into a shared, multilevel institutional framework for thinking about, conducting, and developing research in nanotechnology) is examined for its specific historical contexts and signature events. Of course, nanotechnology has been notoriously difficult to assess. Some have argued that these are fundamentally new technologies that represent a break from past modes of innovation and a catalyst for a new industrial revolution. Other observers point to nanotechnology's roots in various conventional traditions of academic science and see a basic continuity with what came before, suggesting that the way to understand nanotechnology is as a continuation of the steady evolution of scientific knowledge.

Eisler begins this part of the book with an essay (Chapter 2) examining the history of the U.S. National Nanotechnology Initiative (NNI), tracing its origins back to a variety of scientific precursors, policy precedents, and political battles over how science was to be fashioned as a public good in the postwar, pro-science America. Eisler argues that the key to understanding the field of nanotechnology is to recognize that there are indeed significant continuities with the conventional "modernist" trends in the post-1945 politics of basic science. But there are also critical differences that reflect how nanotechnologies are developing in a postindustrial economy that is increasingly dependent on (dis)simulation and speculation. The chapter focuses in particular on the NNI and its broader consequences for the ways academia, industry, and the state collaborate in developing and manufacturing new technological platforms.

For Eisler, nanotechnologies represent a new type of political terrain where basic science is garbed as industrial policy. Eisler starts off by noting that indeterminacy is the defining feature of nanotechnology giving it a "quintessentially postmodern quality." Some see this as word play, but it is more than this: it is the difference between understanding nanotechnology as "a distinct science or engineering practice, and, thus, a coherent body of material and epistemic practices and social relations," or "an umbrella term signifying developments in academic and industrial science over the last quarter century."

Note that Eisler works at the level of discourse. He is interested in how nanotechnology as an idea system of a particular sort took hold and has had consequences for the material world of science made up of agents, practices, laboratories, and so on. He argues that the key to understanding this story is a particular group of people (a broad community of basic science academics) who (discursively) created this thing called nanotechnology. What is it then? What is nanotechnology? It is an idea about how to define the nature of the scientific field that is fundamentally inclusive, focusing on the molecular scale of the object or the process, rather than on some more explicit shared scientific program of research. It is thus, first and foremost, a "big tent" strategy for talking about some constellation of scientific endeavors. It pulls together a broad array of academic scientists as allies and collaborators so that they can act together collectively toward a common goal. And what is that goal? Here Eisler's argument is very clear. They wanted a bigger share of the U.S. federal science budget, and this goal was best accomplished by actively promoting inclusivity.

But, according to Eisler, it is also an idea (a discursive configuration) that matches up with other kinds of ideational or rhetorical matters. Specifically, as Eisler shows, it is an idea about how to talk about basic science research in a way that fits with an agenda for a broader R&D policy, an agenda that highlights the hand-in-glove potential of academic science and economic markets. As he shows, this balance between state-funded basic science and the promotion of market innovations and growth fit well with the rhetorical goals of various constituencies in the field of American politics in the 1990s and aided its articulation into the character and style of the basic national R&D policy. Eisler shows this specifically in the context of a close look (a case example) of the Clinton administration.

And what then has been the result? Eisler tells us that (in some ways) the rhetorical construct fulfilled the ambitions of its proponents. Money did indeed flow toward basic science for the purpose of studying nanotechnology. But the amount of money was less than hoped for, and no new scientific field was generated. Probably, Eisler suggests, this was because "new fundamental physics or theories beyond those of colloid chemistry were not necessarily required." There were certainly some material consequences. A number of projects were funded under the NNI, new professional journals were established, some scientists began identifying themselves as nanoscientists, and so on. But in another sense, there are indications that the moment has passed. A decade has come and gone under this new science regime, and nanotechnology has still had only a minimal impact on our lives, our industries, or our markets. Media attention appears to have peaked (at least for the time being, absent a significant risk event). And meanwhile, a new crisis of the state and the global economy is upon us. A new round of R&D consensus will need to be fashioned. New power struggles will inevitably emerge around the role of the state in the funding and facilitating of scientific practice under these new historical conditions. And so, perhaps, we might conclude along with Eisler, "The political consequences of hyperreality in this case are likely to be minimal."

In Chapter 3, historian Patrick McCray raises questions about how the creative possibilities of science, the sense of where one is going and how one is going to get there, emerge (in part) from the boundary crossing mixture of scientific (and literary) fields. In this provocative chapter, McCray shows how Arthur C. Clarke's 1978 award-winning book about the construction of a space elevator closely mirrored the ongoing bridging of space exploration and nano-fabrication that has brought these two scientific arenas together time and again over the last several decades. The chapter explores the connections between space exploration and nanotechnology as two intertwined thematic frames for promoting the future benefits of science. This linkage (especially at the level of the social imaginary) has been both longstanding and foundational. Thus, Clarke's book, *The Fountains of Paradise*, describes the building of a space elevator using ultra-strong nanomaterials. The book appeared at a time of renewed international interest in space exploration and space colonization. Within a few years, however, pro-space advocates like K. Eric Drexler turned their attention away from space in order to promote the idea of nanotechnologies and molecular manufacturing. This interest in blending the exploration of outer space and nanospace was not limited just to technological visionaries. Prominent nanoscientist Richard Smalley used space applications, especially the space elevator concept, to make the case for greater university investment in carbon nanotube production at Rice University. The chapter explores the reconvergence of space exploration and nanotechnology as witnessed by the interest in space elevator technologies and the engineers and institutions advocating them. It also examines the role of NASA in fostering experimental as well as computation nanotechnology and its place in the larger institutional infrastructure that supports nanoscale research in the United States.

In Chapter 4, Cyrus Mody takes a different tack, showing how the mundane practice of organizing and attending scientific conferences has been a critical factor in generating a sense among scientists and engineers that they shared membership in a discrete and emergent field of scientific work that focused on the nanoscale and bridged many traditional disciplinary boundaries. Mody points out that "[t]he preference of analysts of science, at least since Kuhn, has been to place scientists' explicitly knowledge-producing activities in cultural context . . . Neglected are the mundane activities scientists share with other professionals." To which Mody adds, "Conferences, almost by definition, are where scientists are at their most social, and where competing scientific agendas meet (literally) face-to-face." In the chapter Mody traces the role of conferences as essential to the emergence of "nanoscience" as a legitimate way of describing a specific line of scientific research. Compared to laboratories, field sites, classrooms, and other venues for scientific work, conferences have played a surprisingly impoverished role in the descriptions offered by the social studies of science literature; nor have traditional historians, sociologists, and philosophers of science evinced any greater interest in conferences. This is odd because it is no secret that scientists spend a great deal of their time preparing for and going to a variety of such meetings. Hence, conferences are a promising site in which to catch the "social construction" of scientific knowledge and collaborative enterprise in action. This is perhaps especially the case for an emerging field like nanotechnology where conferences are critical to manufacturing a sense of community and conferring legitimacy on novel terms such as "nanoscience." Starting with the Gordon Research Conferences on Microstructure Fabrication in the mid-1970s this chapter traces the essential role that professional conferences played in assembling a self-consciously defined field of nanotechnology.

The next part of *The Social Life of Nanotechnology* takes up macro-institutional level questions regarding the development of nanotechnologies as a set of scientific fields and considers them from the perspective of the social organization (and political economy) of modern science. American studies scholar Christopher Newfield sets the stage for this part with a broad stroke examination of nanoscale collaboration in relation to the social challenges it attempts to address.

The core question of this chapter is whether nanotechnology as a *social* activity has distinctive features that will allow it to accelerate innovation more effectively than normal scientific practice. The context here is global energy policy. Newfield describes the very real and looming dangers of climate change and argues that nanotechnology is a major source of potential solutions for this problem; but, he warns, it is going to require some remarkable scientific advances to catch up to the speed of climate change events. Indeed, something like Moore's Law (describing frequent doubling in processor speeds) would need to be achieved by nanotechnologies if they are to be productive enough (in energy technology) to generate solutions at the speed that they will be needed for the climate crisis.

In seeking a special feature in the fields and practices that nanotechnology specifically designates, a leading candidate has for some time been thought to be interdisciplinary collaboration. Here, the sense is that the real breakthroughs tend to come from these kinds of collaborations, because these are the developments that really slingshot science forward through the borrowing and redeployment of new ideas, new tools, and new modes of understanding. Whereas most specialists have used striking recent increases in patent and publication quantity to suggest that nanotechnological practice has increased knowledge production, Newfield argues the evidence on this is ambiguous at best, and that both publication counts and their related networks are best regarded as symbols rather than literal signs of scientific and technological progress. Turning to a more direct measure, Newfield and his collaborators conducted a pilot survey of one campus' science community that suggested that collaboration rates are indeed higher among self-identified nanotechnologists. At the same time, responses suggested that collaboration is regarded more as a required baseline of disciplinary scientific practice than as a productivity-enhancing activity. Thus, Newfield argues that it appears that collaborative practices internal to the domain of nanotechnology will not in themselves accelerate the development of solutions of a major challenge like the growing climate emergency. He concludes that rapid nano-enabled progress must start from the premise that nanoscience is normal science in the Kuhnian sense, which will advance in the way that other science does: with supportive, comprehensive social policy whose core elements he briefly outlines.

In Chapter 6, Swedish anthropologist Mikael Johansson refocuses our attention with an ethnographic examination of the changing systems of power and control within the global scientific workplace. Here the question is how nanotechnology has grown up around a new political economy of science in which more traditional expectations for power, privilege, and professional autonomy are being actively reconfigured and redesigned. Johansson points out that nanoscience focuses on atoms and molecules—the last stable building blocks in our physical world, what might be described as nature's "next to nothing." Nanoscientists are (in many ways) an elite social group. The vast majority of nanoscientists come from middle-class backgrounds and, in contrast to the average worker, are highly educated (most have a PhD or they are working to obtain one), highly trained in skilled technologies and working in high value, well-funded organizations. However, most of them work 60–90 hours a week, have no or little vacation time, and many of them are quite poorly paid. In some cases scientists work without healthcare plans and get by on small research stipends. Thus, Johansson tell us that some of the people who study nature's "next to nothing" do this work for "next to nothing." This chapter examines labor conditions within the scientific worlds of nanotechnology, asks why highly educated nanoscientists accept these types of working conditions, and examines how these contradictions reflect the changing character of scientific work in a postmodern, globalizing economy.

In this chapter nanotechnology is used as a case to represent a much bigger set of phenomena that have to do with the changing political economy of scientific knowledge. The political economy includes practical matters, those that operate at the level of a particular individual researcher, a person's—income, vacation time, level of authority, and independence in the labor process, as well as a breadth of other working conditions more generally. But of course, these are not arbitrary changes. At other levels of social organization, broader social processes and trends can be pointed to as causes (and dually, as consequences) of the more micro-level phenomena that are highlighted here. So, for example, location in the scientific world system (moving from a position in the scientific periphery to the core) contributes to the less dominant (disempowered) situations of some of these nanoscience workers. Another issue Johansson points to is the changing character of the academic scientist career system. As he indicates, one of the profound changes in American academia (and elsewhere as well) is that there has been a shift away from the postwar expansion of the professoriate that had led to the rise of a broader, more open, and, in a sense, a more democratically organized academic system. For the last decade or two that system has been eroding, thanks to a shrinking of resources and significant limitations in the growth of hiring of tenure-track faculty along with a concomitant increase in the number of PhDs who work in the academic field under very different labor (and career) conditions. One familiar example is the rise of itinerant lecturers who are not offered tenure at any institution but who are increasingly being pressed into service to teach the undergraduate curriculum. Johansson highlights a parallel trend linked to the expansion of two-track careers among tenured faculty and the growing cadres of otherwise employed career scientists (and technicians) in nanotechnology laboratories.

Other issues are also salient here. As the higher education system is plunged ever deeper into a fiscal crisis, so are new types of funding arrangements being tested out that link academia and industry together in a variety of new (and more expansive) collaborative combinations (and revenue generating arrangements). This has led to the growth of alternative sources of research funding as well as new systems of reward for academics who operate for-profit spinoff firms to capitalize on their research findings. These developments are also changing the political economic character of the field, and the view of the scientific enterprise by lay publics. These are big issues reflecting enormous international changes in the nature of academic science. A real strength of Johansson's paper is that he shows how nanotechnology embodies many of these emerging trends.

In Chapter 7, Richard Appelbaum, Cong Cao, Rachel Parker, and Yasuyuki Motoyama compare national nanotechnology policies in the United States and China, examining the degree to which their differing approaches constitute industrial policies, in the sense of, for example, directing public funding toward commercialization rather than basic research. There is considerable contrast between these countries in the degree to which government direction of economic development—so-called "placing bets on specific industries"—is politically acceptable; yet both countries have, to varying degrees and through differing processes, "placed bets" on nanotechnology as a key driver of commercialization and reindustrialization. Through a comparative study based on interviews, bibliometric analysis, and the analysis of patent data, the chapter assesses the degree to which each country has an industrial policy approach to the commercialization of nanotechnology, and the degree to which such efforts have resulted in the successful move from lab to product.

Appelbaum et al.'s main point is that in contrast to "the conventional understanding of the state-market relationship in the United States in which technology is developed in a bottom-up approach by the market, in lieu of the top-down approach by the federal government," they find that nanotechnology actually did come from the top down, in the sense that it was not market actors but agents of the state who were responsible for initiating and indeed shepherding into existence the broad field of activities of nanotechnology (or at least this is true insofar as the birth of the NNI is concerned).

In Chapter 8, sociologists Rachel Parker and Richard Appelbaum examine some implications of China's move to high-tech innovation for U.S. policy. They note that the Chinese economy has been tied to the United States for over two decades. But now—thanks to the $2 trillion dollars (or so) in foreign reserves that China has amassed through export-oriented industrialization—the country is seeking to become an "innovative society" by "leapfrogging development," by investing vast sums in developing an innovative and globally competitive R&D capability. Nanotechnology is one of the key areas selected for investment; energy is a central focus of China's efforts in general. China by now has developed a large and growing internal market, and many economists now claim that the Chinese economy is too investment driven and insufficiently consumption driven. Parker and Appelbaum argue that China is now poised to switch from export-oriented industrialization (EOI) to import-substitution industrialization (ISI), which in turn means that it no longer sees its fate as coupled tightly with that of the United States as it has in the past: it no longer requires the United States, or other foreign consumers, as the principal driver of economic growth. Although China cannot disconnect its economy from that of the United States precipitously without jeopardizing the value of its dollar reserves, Parker and Appelbaum argue that in the long run just such an uncoupling is likely to occur, as China moves up the value chain—increasingly designing and marketing its own high-technology products, selling to its own growing internal market, and increasingly offshoring its manufacturing to Vietnam and low-wage countries in Africa. The chapter concludes with a discussion of the policy implications of these developments—specifically, how the United States and other countries can partner with a resurgent China, rather than resorting to protectionism and a renewal of great power geopolitics.

The last part of the volume focuses in more closely on the negotiation between various institutional agents (including social science researchers) and citizens over what counts as valid knowledge, what are the forms and styles of truth, and the potential for benefits, risks, or harms. In particular, nanotechnologies in the publics' views reflect their risk status as a "new species of trouble" (Erikson, 1995), highlighting the (im)possibilities of independently verifiable knowledge in the world of molecular scale technologies. Scholars in the fields of risk perception and communication make a special contribution here as they strive to work upstream, staying ahead of the institutional curve, and eliciting constructed preferences as the very meaning of nanotechnologies as enabling technologies with a particular set of ecological and social footprints is being carefully and intensively negotiated and, frequently, contested. Novel upstream research and engagement efforts challenge publics and experts to anticipate feelings, judgments, and actions for whole new classes of technologies, and to imagine them as active agents in social contexts that may reproduce, exacerbate, or ameliorate current inequalities, recreancy concerns (Freudenburg, 1993), or obstacles to democratic institutions. Some sectors of the public are readily seen as "implicated actors"—that is, those in the social world of nanotech R&D who "are silenced" or are "only discursively present" (Clarke, 2006, p. 46). Here the very questions of the expert(s), the public(s), their respective roles and interests, as well as their capacities for being engaged in an informed technological debate, are precisely what is at stake. Chapters consider how broader (mass media news) frames, cultural logics, preexisting scientific narratives, and professional intellectual frameworks have played a role in helping to shape the social imaginary as it applies to the new and emergent field of nanotechnologies.

This concept of upstream engagement is now firmly embedded in social researchers' and policymakers' vocabularies—but the question of *how* to legitimately dialogue with the public on emerging technologies remains pertinent. In common with other upstream technologies, nanotechnology's challenges in this regard are fraught with conceptual and practical difficulties. In Chapter 9, Adam Corner and Nick Pidgeon review the full range of completed public engagement work on nanotechnologies and ask whether there is a model that demonstrably "works." They discuss this work in relation to previous attempts to involve the public in scientific matters, referring specifically to the "GM Nation" project in the UK. They also outline some of the lessons that nanotechnology provides about how to approach the social and ethical questions raised by other emerging technologies (such as synthetic biology, parts of climate geoengineering using nanoparticles, etc.). Corner and Pidgeon conclude by arguing that despite the multiple challenges of designing and implementing upstream engagement programs, it is possible to identify approaches that work—and that these models will continue to serve an important function in broadening the scope of public involvement in decision making about science and technologies.

In Chapter 10, Jennifer Rogers, Christine Shearer, Barbara Herr Harthorn, and Tyronne Martin argue that the emergence of nanotechnologies provides a special opportunity to examine public perceptions of technological risk and benefit as they are being socially and culturally produced. They write, "[E]merging technologies such as nanotechnologies provide a unique opportunity to examine perceptions of technological risk and benefit as they are being socially and culturally produced, rather than retroactively." The chapter examines how the application context within which nanotechnologies are deployed shapes public perceptions about the risks and benefits of the science, and considers why this is so. Drawing on data collected from a series of six deliberative workshops conducted in California in 2009, and building on their prior comparative U.S.–UK deliberations in 2007, the chapter argues that the particular industry or application context is critical in shaping public responses to nanotechnologies, and that these "application effects" are most evident around axes of urgency/necessity, novelty, regulation, distribution of benefits and risks, privacy, and responsibility. The chapter also provides a critique of the uses of public deliberation to enhance public participation in technological decision making, to intervene in the new public deficit—"engagement deficit"—that is argued by some to have supplanted the public "knowledge deficit" of prior decades. In particular the issues surrounding introduction of European-style engagement models in the upstream/midstream in the United States are examined.

The authors argue that as participants interpreted workshop discussion points and materials they drew on their understanding, values, and past experiences regarding health or environment domains more broadly, understandings and values that often varied by technological application context or industry. Such variables include cultural knowledge about environmental and energy concerns, technological innovations, and the health industry, as well as wider social issues such as inequality and trust in government. One of the most important and salient findings was that people are worried about global warming and the looming environmental crisis in a way that fully conditions their willingness to take a risk on nanotechnology because if the world is already at great risk, then, what's a "little" more risk, so long as it holds out the potential for fixing the overarching crisis? Another clear and consistent finding seems to be that people (some people anyway), even in the techno-optimistic United States, are wary about the "everything will be fine" attitude that characterizes so much of the modernist embrace of scientific progress. In a sense that follows one of the main themes of the edited volume, there is a healthy postmodern skepticism evident here, a sense of cynicism that makes many respondents presume that whatever benefits (and risks) might be derived from these technological advances will likely be unevenly (and unfairly) distributed.

In Chapter 11, electrical engineer Erica Lively teams up with political scientists Meredith Conroy, David Weaver, and Bruce Bimber to argue that publics learn about nano applications and their potential implications primarily through media coverage—traditional print media as well as new media. The nanotechnology case is unusual in that journalists have rarely confronted an entirely new domain of science and technology in such a short segment of time. In the United States, the development of the NNI presented journalists with a novel issue, and the chapter explores how journalists responded to these developments by examining two key aspects of news coverage from 1999 to 2009: the overall volume of English-language print media coverage and the framing of newsworthy elements. The study demonstrates the patterned rise and fall of nano-centric news content over this 10-year period and examines how journalists applied frames involving progress, risk, and regulation to make sense of developments in nanoscale science and technology. The analysis focuses on comparisons among major U.S. news outlets.

In Chapter 12, environmental sociologists William R. Freudenburg and Mary B. Collins argue that most research to date has pointed out that although public awareness and knowledge of nanotechnology is low, when prompted those surveyed report positive opinions. Given this, it has been difficult for researchers to document substantive opinion characteristics, make potential recommendations for the future, and provide general assessments—especially in the context of perceived risk. Instead, what this climate *does* offer is an excellent opportunity to test Freudenburg's recreancy perspective. Recreancy is defined by Freudenburg (2000) as "the failure of experts or specialized organizations to execute properly responsibilities to the broader collectivity with which they have been implicitly or explicitly entrusted" (p. 116). The recreancy perspective grew out of analysis of past technological controversies. Freudenburg has noted that it has become commonplace to "blame the public" during these periods of controversy. This blame is situated in a framework that assumes that low levels of knowledge (or of sophistication) cause the public to be confused as to the nature of the controversy. In this case, usual remedies call for widespread public education campaigns. The recreancy perspective challenges this framework and shows a distinctive lack of evidentiary support for "blame the public" style arguments. Instead, the recreancy perspective posits that there is a fundamental misunderstanding about what it means to live in a technologically advanced society in the twenty-first century. According to Freudenburg and Collins, little change in public knowledge/opinion is expected until and unless one or more "signal" events occur. Given the low levels of nanotechnology awareness and substantial amounts of scientific uncertainty (related to both societal benefits and risks), the nanotechnology context provides an ideal case to study the recreancy perspective via empirical testing and testable hypothesis exploration. In particular, this work raises issues directly relevant to the "responsible development" position mapped out by NNI leaders—that is, that the organizations and institutions leading the R&D effort for new technologies bear a responsibility for institutional integrity, for *trustworthiness*, that includes attention to the legitimate concerns of the citizenry, the taxpayers whose largess funds the public sector development investments, and who ultimately have the right to question the promises as well as the perils involved in technological development. This work dovetails neatly with the work of science and technology studies scholars such as Sheila Jasanoff, who in her book *Designs on Nature* (2005) has looked at the institutional apparatuses of different states, and how they produce distinct "civic epistemologies" about decisions on science and technology.

In sum, this volume speaks to a broader understanding of technoscientific fields and their connections to institutional and market systems as well as individual level behavior and thought. The analyses reported in the volume ask what careful and systematic investigations of the social life of nanotechnology can tell us about future prospects and likely trajectories for this field of science and engineering and for global societies that hope to ride the wave of economic growth and prosperity on which their funding is premised. And, more generally, the authors ask what the role of the social sciences and humanities can and should be in relation to emerging fields of science such as nanotechnology.

ACKNOWLEDGMENTS

This material is based upon work supported by the National Science Foundation under Cooperative Agreements Nos. SES 0531184 and SES 0938099 to the Center for Nanotechnology in Society at University of California at Santa Barbara. It has also received support from the National Science Foundation and Environmental Protection Agency under Cooperative Agreement No. DBI 0830117 to the UC Center for Environmental Implications of Nanotechnology at UCLA and University of California at Santa Barbara. Any opinions, findings, and conclusions or recommendations expressed in this material are those of the authors and do not necessarily reflect the views of the NSF or EPA. This material has not been subjected to EPA review and no official endorsement should be inferred.

REFERENCES

Appadurai, A. (Ed.). (1986). *The social life of things: Commodities in cultural perspective*. Cambridge: Cambridge University Press.

Beck, U. (1992a). From industrial society to the risk society: Questions for survival, social structure, and ecological enlightenment. *Theory, Culture, and Society, 9*, 97–123.

Beck, U. (1992b). *Risk society: Toward a new modernity*. Newbury Park, CA: Sage.

Beck, U., Giddens, A., & Lash, S. (1994). *Reflexive modernization: Politics, tradition aesthetics in the modern social order*. Cambridge: Polity Press.

Bowker, G., & Star, S. (1999). *Sorting things out: Classification and its consequences*. Cambridge, MA: MIT Press.
Bourdieu, P. (1984). *Distinction: A social critique of the judgement of taste*. Cambridge, MA: Harvard University Press.
Brown, J. S., & Duguid, P. (2000). *The social life of information*. Cambridge, MA: Harvard University Press.
Cetina, K. K. (1991). Epistemic cultures: Forms of reason in science. *History of Political Economy, 23*(1), 105–122.
Cetina, K. K. (1999). *Epistemic cultures: How the sciences make knowledge*. Cambridge, MA: Harvard University Press.
Clarke, A. E. (2005). *Situational analyses: Grounded theory after the postmodern turn*. Thousand Oaks, CA: Sage Publications.
Erikson, K. (1995). *A new species of trouble: The human experience of modern disasters*. New York: W.W. Norton & Co.
Freudenburg, W. R. (1993). Risk and recreancy: Weber, the division of labor, and the rationality of risk perceptions. *Social Forces, 71*(4), 909–932.
Freudenburg, W. R. (2000). The "Risk Society" reconsidered: Recreancy, the division of labor, and risks to the social fabric. In M. J. Cohen (Ed.), *Risk in the modern age: Social theory, science and environmental decision making*, pp. 107–122. New York: Palgrave.
Funtowicz, S., & Ravetz, J. R. (1992). Three types of risk assessment and the emergence of post- normal science. In S. Krimsky & D. Golding (Eds.), *Social theories of risk*, pp. 251–273. Westport, CT: Greenwood.
Funtowicz, S., & Ravetz, J. (2003, February). Post-normal science. International Society for Ecological Economics. *Internet Encyclopaedia of Ecological Economics*, pp. 1–10. Retrieved November 26, 2011, from www.ecoeco.org/pdf/pstnormsc.pdf
Giddens, A. (1990). *The consequences of modernity*. Cambridge: Polity Press.
Jasanoff, S. (2005). *Designs on nature: Science and democracy in Europe and the United States*. Princeton, NJ: Princeton University Press.
Kopytoff, I. (1986). The cultural biology of things. In A. Appadurai (Ed.), *The social life of things*, pp. 64–94. Cambridge: Cambridge University Press.
Latour, B. (2005). *Reassembling the social: An introduction to actor–network-theory*. Oxford: Oxford University Press.
Ravetz, J. (1999). What is post-normal science? *Futures, 31*, 647–653.
Rogers-Hayden, T. A., Mohr, J., & Pidgeon, N. (2007). Introduction: Engaging with nanotechnologies—engaging differently? *NanoEthics, 1*(2), 123–130.
Rowe, G., & Frewer, L. (2000). Public participation methods: A framework for evaluation. *Science, Technology, & Human Values, 25*, 3–29.
Stilgoe, J. (2003). *Citizen science*. London: Demos.
Star, S., & Griesemer, J. (1989). Institutional ecology, "translations" and boundary objects: Amateurs and professionals in Berkeley's Museum of Vertebrate Zoology, 1907–39. *Social Studies of Science, 19*(3), 387–420.
Weiner, A. (1992). *Inalienable possessions: The paradox of keeping while giving*. Berkeley: University of California Press.
Whitmarsh, L. (2009). Review of Dietz and Stern, *Public participation in environmental assessment and decision making. Environmental Science & Policy, 12*, 1069–1072.
Wynne, B. (1992). Misunderstood misunderstanding: Social identities and public uptake of science. *Public Understanding of Science, 1*, 281–304.

Part I

Constructing the Field of Nanotechnology

The Social Origins of Nanotechnology

2 Science That Pays for Itself

Nanotechnology and the Discourse of Science Policy Reform

Matthew N. Eisler

Nanotechnology's indeterminacy has haunted scholars of science, technology, and society, if not policymakers, the public, and tellingly, many physical scientists, since the term "nanotechnology" began to enter popular usage more than two decades ago. Relatively few have shown a willingness to come to semiotic and syntactic grips with the word. A "standard opening gambit" of nanotechnology papers in the 2000s, noted Johnson (2009, p. 144), was simply to recapitulate the authoritative definition of the National Nanotechnology Initiative (NNI): the science and engineering of objects between 1 and 100 nanometers. This, she held, failed to clarify the material identity of nanotechnology and, perhaps even more importantly, the identity of nanotechnologists.

Nevertheless, most social scientists would agree that the appearance of a major federal program in nanotechnology's name in 2000 stemmed from a debate arising at the end of the Cold War concerning the role of undirected basic science as a productive force in the broader American economy. But there is no consensus on whether nanotechnology and the NNI broke with trends in U.S. science policy. Some have argued that the initiative was born of industrial policy (Block, 2008) or itself represented a form of industrial policy (Motoyama, Appelbaum, & Parker, 2011). Others have leaned toward continuity. Johnson (2004) held that the initiative was the centerpiece of a White House science policy "fundamentally" different from its predecessors, yet saw nanotechnology as the latest effort to rationalize basic science for its practical utility. McCray (2005) observed that the backers of what became the NNI used well-tried justifications for state-backed basic science, combining appeals to international competitiveness with the proven tactics of broad-front alliance-building. Mowery (2011) did not believe the NNI embodied fundamentally new socio-institutional relations, noting that cross-institutional collaborative research and development involving academia, industry, and government occurred throughout the post-1945 period. Innovation in nanotechnology, he added, was characterized by "vertical specialization" (the distribution of innovation among firms involved in developing commercial technology), a term associated with the information and biological technology industries. Mowery understood the NNI as an organization of directed research and development. What was novel, he posited, was that, unlike in the IT and biotechnology sectors, patenting in the nanotechnology sector increasingly covered fundamental advances at very early stages of research (p.708).

How to reconcile these positions? Few students of the subject would dispute that nano-prefixed technical terminology became associated with "nanotechnology," a neologism with origins in academic materials research but popularized and invested with utopian overtones in the 1980s by futurists, most notably K. Eric Drexler. Some have interpreted this as a discourse of hyper-reality, adopting Baudrillard's term for the blurring of distinctions between the real and unreal in an era of powerful technologies and politics of simulation (Milburn, 2008). Former National Science Foundation (NSF) director Neal Lane (2001) claimed that the Nobel laureate chemist Richard E. Smalley understood nanotechnology as a compound word expressing both current basic research (nano) and deep-future application (technology). Nordmann (2009) held that matter-machine metaphysics informed stories portraying nanotechnology as a form of science that proceeded "in an engineering mode" (p. 125).

In a similar vein, researchers have dispatched what Rip (2006) referred to as theories of the "folk sociology" (unverified generalizations and terminologies purporting to explain the world that were oriented to future action in ways that served the purposes of their inventors) of nanotechnology (p. 349). They problematized the "origin stories" that traced nanotechnology to the development of scanning/tunneling/atomic force microscopy in the 1980s or to a lecture delivered by the physicist Richard P. Feynman at Caltech in 1959. Pointing out the iterative nature of progress in science and technology, social scientists have investigated practitioners who studied, modeled, or manipulated materials on the nanoscale prior to the advent of nano-prefixed terminology (Choi & Mody, 2009; Johnson, 2009; Kim, 2008; McCray, 2009; Mody, 2004, 2006; Schummer, 2006).[1]

Scholars have offered a number of explanations of the ends of the authors of folk theories of nanotechnology. These explanations have ranged from efforts to broaden basic science constituencies through institution-building and expansion of service infrastructure (Mody, 2009) to justifying specific projects (Rip, 2006) or institutional missions of basic science at a time when the federal basic science establishment had come under strong political scrutiny (Gallo, 2009; McCray, 2005). These ideas are not necessarily mutually exclusive. What is often lacking in accounts of the historical development of nanotechnology and the NNI, however, is explicit consideration of how the construction of folk theories related to the political economy of the U.S. federal basic science establishment.

Kleinman's (1995) observations of the contours of the science policy landscape in the early and mid-1990s provide helpful context. Writing in reference to the NSF, he echoed Smith (1990) in noting the remarkable stability of the federal science establishment once it crystallized in the early postwar years. Fragmented and resistant to centralized planning, the institutions of federal science mirrored the political structure, where power was distributed and parties did not adhere to platforms or enforce discipline. In such a system, the scope of institutional change was limited. Creating powerful new agencies was difficult; it was easier to make incremental changes to existing ones or eliminate small ones altogether (see Bimber, 1996). Far more mutable, held Kleinman (1995), was "symbolic capital." As the state fiscally retrenched, the status of high-tech industry ascended, and federal science entered a period of crisis in the early 1990s, he suspected that a change of science policy discourse was imminent. The waning capacity of the traditional symbolic capital of basic science as the "bedrock of economic well-being" to effect organizational change in the interests of scientists, he predicted, might lead them to associate their work with industrial technology (pp. 173, 193–195).

Nanotechnology rhetoric began to appear in science policy discourse several years after Kleinman (1995) published this observation. Of course, institutions of federal science had been subject to scrutiny from public officials concerned that resources be utilized to address a range of pressing social issues since at least mid-1960s. Over time, reforms were put in place to monitor and control the priorities, integrity, and productivity of state-supported basic science. From 1980, Congress passed a series of laws designed to ease the transfer of publicly funded research performed in universities and federal laboratories to the private sector, from whence, reasoned politicians, it would be transformed into practical, wealth-generating technology. These reforms have been interpreted as a watershed in the social relations of basic science (Forman, 2007; Guston, 2000; Mirowski, 2004, 2011; Rosenberg, 1994; Ziman, 2000).

But in the wake of end of the Cold War, the political masters of the federal science establishment continued to perceive that it contributed little to economic growth at a time when the country was slowly emerging from recession. To be sure, few questioned the fundamental premise that had justified the construction of this institution after the Second World War, that national programs of basic science begot technology that somehow served national goals (frequently referred to as the linear model of innovation). Debate instead turned on how to further reform federal science so that it would yield the desired results, a discussion that frequently lurched into invective and threats to eliminate certain federal agencies altogether.

The more radical ideas were never implemented. But as federal science policy discourse increasingly suggested that basic science had to pay a technological dividend, representatives of state-dependent science communities believed that thoroughgoing reform was imminent. One of their responses, I argue, was to conceive nanotechnology, with its matter-as-machine metaphysics, as a kind of symbolic capital that would help defend their interests by rhetorically aligning them with lawmakers.

Accordingly, a more satisfactory perspective on nanotechnology's identity can be gained by understanding nanotechnology as a way of talking about basic physical research and its economic potential that, for a time, became an important element of science policy discourse. Bazerman (1988) observed that symbolic representations in scientific writing have significant material effects, with the discourse they support tending to be hidden within what appears to be the transparent expression of a fact. And the process of ascertaining "facticity" is always guided by human values and interests. As Harris (1997) points out, scientists routinely deploy rhetoric as they "argue in the making of knowledge" (p. xii). Excavating nanotechnology's role in science policy discourse reveals that scientists and their supporters mobilized the term, and its connotation of practical macroscale technology, to resolve a dispute over knowledge that was primarily social, not physical, in nature: the presumed economic value of state-supported basic research. This rhetoric, I argue, comprised the master folk theory of the many invoked by observers of and participants in the nanotechnology project.

In the first part of this chapter, I outline how Drexlerian historical philosophy, progress in certain fields of science and technology, and the vicissitudes of federal science politics helped embed "nanotechnology" within federal science policy discourse. In the second part, I briefly assess the results of the nanotechnology enterprise. As a social phenomenon, I argue, nanotechnology should not be regarded as cause or consequence of a historical saltation in science and technology or federal science, technology, and industrial policy. It should instead be seen as a product of trends in the federal basic science establishment during a periodic reckoning with the broader American polity.

(NANO)TECHNOLOGY AS RESEARCH AND DEVELOPMENT POLICY

Analyses of the precepts of nanotechnology often refer to the role of K. Eric Drexler. A familiar figure to specialists in contemporary studies of science, technology, and society, he is usually interpreted as one of the chief popularizers of the term "nanotechnology," a word coined in 1974 by the electronics engineer Norio Taniguchi (1974) to refer to a hypothetical materials finishing process based on the application of advanced versions of existing microfabrication technologies, especially ion sputter-machining. What scholars of science, technology, and society have dwelt less on is Drexler's understanding of the social relations of basic science in the 1980s and 1990s. His chief achievement, I argue, was to craft novel metaphors that reinforced assumptions of the economic potential of basic science held by Congress and presidential administrations in this period.

An aerospace engineer by training, Drexler worked as neither an experimentalist nor a theorist, properly speaking. Relying heavily on computer simulation, he blended imagery of classical macroscale machines, nanoscale science, and evolutionary biology in a way that evoked precision and total control (Bensaude-Vincent, 2004).[2] Drexler wove these motifs into a historical philosophy based on inductive inference of the laws of matter on the physical, biological, and social levels. For him, the natural and social worlds were one. Technological progress was socio-biological. As nature evolved, so, too, did technology. The co-evolution of society and nature presaged the emergence of what Drexler referred to as the molecular assembler. A hypothetical chemical analogue possessing the qualities of a self-assembling protein molecule but utilizing more robust materials, he imagined this as an immensely powerful programmable form of biotechnology capable of building any macroscale artifact atom-by-atom with no waste but also causing something like a pandemic—the infamous "grey goo"—if not carefully controlled by engineers (Drexler, 1986).

Of course, Drexler was only the latest to analogize life as machine. But his ideas reflected certain values and attitudes held by lawmakers, bureaucrats, and industrialists in the 1980s and 1990s. Drexler's belief that assemblers would one day render social policy irrelevant by creating new wealth from the limitless cornucopia of the cosmos channeled Reaganite supply-side economics, as well as the technocratic tradition and deep-seated beliefs in American culture of the place of the individual in science, technology, and society. Drexler's promise of a transdisciplinary science field of vast technological potential also paralleled the Reagan administration's claim that science was the basis of economic growth and valorization of collaborative relations between scientists and technologists (see Reagan, 1985).

It is instructive to compare and contrast the connotations summoned by Drexlerian nanotechnology discourse and by nano-prefixed terminology associated with the new microscopy and microfabrication technologies that began to appear in the 1980s. Mody (2009) observed how U.S. academic researchers and science equipment suppliers came to appreciate the strategic potential of language in advancing their respective material interests, altering terminology to express progress and accrue symbolic and real capital. Within microfabrication communities, he held, existing nomenclature (the micron) was perfectly adequate for describing phenomena at the nanoscale (Mody, 2009). Encouraged by instrument manufacturers eager to promote scanning tunneling/atomic force microscopy, researchers conjugated terms like "nanoengineering," "nanofabrication," and "nanostructure fabrication" as indices of their ability to characterize and manipulate matter on an ever-smaller scale. They also adopted this lexicon in order to broaden disciplinary participation in academic research conferences as corporate involvement in these gatherings diminished from the late 1970s (2004, pp. 129–130).

Drexlerian nanotechnology discourse was used to serve similar roles but cast a far wider ontological net. The Rice University chemist Richard Smalley was one of the first to appreciate its application in science politics and became one of its highest-profile interpreters. Famous by the early 1990s as one of the discoverers of C_{60}, the first fullerene, Smalley believed Drexlerian "nanotechnology" was useful primarily as a communication expression. He outlined his thinking in a letter to the deans of engineering and natural sciences at Rice. The chemist had been disappointed that a conference on "Science at the Atomic Scale" held in Japan in January 1991 and sponsored by the journal *Nature* had been dominated by those who "saw nanotechnology as an evolution of microelectronics." Given that chemistry dealt with nanostructures and that biology was the "real nanotechnology of living cells," held Smalley, he was surprised that the event attracted few chemists and biologists. Holding that the "central idea of nanotechnology"—the purposeful manipulation of matter on the nanometer scale—applied or would shortly apply to a large number of fields, he concluded that the word should be interpreted much more broadly. When it was, he claimed, interest among scientists would greatly increase (Smalley, 1993).

Smalley then pondered the rhetorical uses of nanotechnology strictly in relation to his own professional ambitions at Rice, where he was becoming interested in the research, development, and production of carbon nanotubes (Mody, 2010). But later in the decade, representatives of science communities increasingly began to invoke the word in the sense Smalley mooted, as a way of connoting interdisciplinary science with practical implications. This occurred at a time of unprecedented discord in federal science. For researchers dependent on state patronage, these were worrying times. Scientists were apprehensive of the Clinton administration's plan to support advanced technology as a major instrument of social policy (Clinton & Gore, 1993), thinking this might divert resources from basic science (Greenberg, 2001). Worse, both national political parties were subjecting the premises of basic research to new scrutiny, matching heated rhetoric in the legislature with action. In October 1993, the Democratic-dominated legislature cancelled the Superconducting Super Collider, the holy grail of particle accelerators. The seizure of both houses of Congress by the Republican Party in the mid-term elections of 1994 further stoked tensions. Conservatives attacked Clinton's technology programs as corporate welfare and focused new scrutiny on federal science agencies. As they struggled to shape the budget for fiscal year 1996, representatives from both sides of the aisle called for trimming the federal science bureaucracy.[3]

Uncertainty and confusion reigned in physical science communities. Justifying the benefits of health research in Washington was "not a hard sell," as Greenberg (2001) noted. Defending undirected basic physical science was less straightforward. But the partial shutdown of the government in late 1995 and early 1996 as a result of the failure to pass the budget bill provided a rallying point for concerted political action among physical scientists that would open space for nanotechnology advocacy. In a June address to a gathering of the American Association for the Advancement of Science, NSF Director Neal Lane framed the budget battle as an attack on science.

Implying that the Republican approach of selective cuts aimed to divide the community, he called for solidarity. A physicist by training, Lane urged scientists to jump into the policy arena, emphasizing their responsibility in educating the public on how science served the "nation and its citizenry".[4] Irritated that his agency had been shut down during the budget crisis and critical of the "perceived stony silence" of scientists during that crisis, Lane maintained in early 1996 that "this year cannot be business as usual".[5]

It was around this time that Mihail C. Roco, a professor of mechanical engineering at the University of Kentucky and a civil servant in the NSF's Engineering Directorate, joined with a small group of supporters in launching what became the interagency nanotechnology movement. To be sure, the NSF had long defended the premise of linear innovation, or, more precisely, the link between national programs of basic research, technology development, and national economic growth. In 1968, it sponsored research contesting a Pentagon study that found only a tenuous connection between basic research and practical technology (Mirowski, 2011). From 1972, the National Science Board, the NSF's governing body, more or less explicitly reiterated the economic value of basic science in its biennial "Science Indicators" (renamed "Science & Engineering Indicators" in 1987). But it tended to qualify this claim, noting the difficulty of assessing science by conventional economic standards owing to its public nature, the time lag in application, and the fact the originators of research did not necessarily reap the benefits (e.g., see NSB, 1996, pp. 8.1–8.11).

In contrast, the interagency group framed the presumed social relations of basic science in unambiguous terms. In a series of studies conducted in the late 1990s, it asserted the reality of a direct connection between academic research and industrial technology, framing nanotechnology as the link. When associated with "nanostructure science and technology," nanotechnology spanned all the conventional divisions of the innovative process, encompassing basic science, esoteric materials, applied science, engineering, and commercial macroscale technologies, especially pharmaceuticals and certain specialized electronics (Siegel et al., 1999). The scale and complexity of nanotechnology, held the interagency group, meant that manufacturing could not occur without basic science developed by the federal government. Once this happened, industry would supplant the state as the chief investor and "nanotechnology" would have the potential to trigger an industrial revolution. Crucially, the interagency group argued that nanotechnology already existed in the form of certain lines of materials research supported by the federal government and even certain consumer products (IWGN, 1999).

By 1999, Smalley was working with the interagency group in lobbying for an expanded federal effort in "nanotechnology." Testifying in Congress in June, Smalley blended Drexlerian metaphysics with small-r republican sensibilities. He moved seamlessly between nanotechnology as both a biological phenomenon (a product of evolution) and a social phenomenon, a "small science" with broad applicability in medical, power source, and transportation technologies (McCray, 2005). Smalley cited drugs, fuel cells, and, in a nod to the Clinton administration's Partnership for a New Generation of Vehicles, a collaborative program of research and development with Detroit, an all-electric supercar (U.S. House of Representatives, 1999).

Interestingly, the interagency group itself did not reflect the idealized social relations of collaborative research and development it claimed were necessitated by nanotechnology. With the exception of Hewlett-Packard scientist Richard Stanley Williams, all were academics by profession with strong connections with the federal science establishment. However, several high-profile figures with links to the electronics industry did lend important support to the idea of nanotechnology.[6] Williams was the interagency group's industry representative. In late 1999, Yale University's Mark Reed, a former employee of Texas Instruments and the discoverer of quantum dots, all but proclaimed an imminent revolution in computing technology, remarking that "dirt cheap" molecular electronics, an obscure field dedicated to replacing silicon as a semiconductor material with alternate substances, would create a "discontinuity." These statements, reported the *New York Times*, impressed the Clinton administration as it planned a national nanotechnology initiative (Markoff, 1999). And Gordon Moore introduced Clinton in an event at the California Institute of Technology in January 2000 that saw the president deliver a speech launching the NNI. One of Caltech's most illustrious alumni and an archetypal entrepreneur of the information age, Moore did not mention nanotechnology at all, offering instead a paean to the presumed social relations of science. Federally supported academic research, held the Intel co-founder, bred commercial, and hence, national prosperity.[7] Clinton then elaborated this theme at length. The occasion was not devoted exclusively to the NNI, as is often assumed. Instead, the president announced the initiative as just one part—$500 million—of a $2.8 billion increase in research funding, money the president emphasized was to be divided amongst *all* the fields of basic science (Clinton, 2000).

SCIENCE AND THE STATUS QUO

What goals were served by introducing the NNI in this manner at this time? At a time when Congress opposed incentive packages for the White House's major technology enterprises but not necessarily their attendant basic science programs, as with the Climate Change Technology Initiative,[8] the Clinton administration seems to have understood the NNI as a form of rhetorical *realpolitik* that reconciled its ambitious technology agenda with Capitol Hill politics.

Lane suggested the NNI helped redress the long-standing power imbalance between the NIH, long the largest federal patron of science, and the DOE, DOD, NASA, and the NSF (Lane, 2001; Lane & Kalil, 2005). The director of the Department of Energy's Office of Basic Energy Sciences succinctly expressed this thinking in an advisory committee meeting in November 1999. The NIH, observed Patricia Dehmer, had required no high-profile initiative to induce Congress to nearly double its budget in the late 1990s, because politicians were convinced this investment would yield significant results. Similar largesse for the physical sciences, she held, hinged on the NNI's "promise of things yet unimagined".[9] Managers like Dehmer took direction from Lane and Kalil, who prompted them to couch their budget requests in terms of nanoscale science and engineering in exchange for a promise of support (Lane & Kalil, 2005).

But the NNI did not herald a shift either in social or physical paradigms. A decentralized coordinating mechanism fitting seamlessly into the federal science polity, the NNI had no fiscal authority (National Research Council, 2009). It was not a funding agency. Its budget was the sum of the contributions of federal R&D agencies, which had the latitude to address the initiative's priorities as a function of their own missions. Moreover, annual federal spending on basic research had been steadily rising well before the advent of the NNI, nearly doubling between 1990 and 2008 from around $17 billion to over $32 billion in constant 2000 dollars. Funds allocated under the rubric of the NNI comprised a tiny fraction of the $248 billion the federal government invested in basic research in this period (NSB, 2010, Appendix Table 4–8). Perhaps most importantly, where the interests of physical sciences communities were concerned, the NNI did not significantly alter the pattern of distribution of federal funding. The NIH continued to dominate as it always had, expanding its share of this sector from about 59 percent to over 65 percent between 2001 and 2009 at the expense of every other major federal player except for the NSF, whose share grew slightly in this period (NSB, 2010, Appendix Table 5–3).

To be sure, the NNI did help justify the formation of many user facilities after 2000, including the 14 university-based laboratories of the NSF-funded National Nanotechnology Infrastructure Network and the Department of Energy's five Nanoscale Science Research Centers. So, too, did dedicated peer-reviewed publications in nanoscale science and engineering proliferate in this period, although, tellingly, most publications used nano-prefixed terms in their titles, not the loaded keyword "nanotechnology," with its utopian connotations.[10] By the end of the decade, a few U.S. postsecondary institutions, all relatively small and peripheral, offered nano degrees, although, again, the word "nanotechnology" was rarely used.[11] Nanotechnology had not yet been established as a true discipline at the level of academic science, likely because new fundamental physics or theories beyond those of colloid chemistry were not necessarily required to describe the interaction between particles in the 1–100 nanometer range and other substances in the environment (Auffan, Rose, Bottero, Lowry, Jolivet, & Wiesner, 2009).

Nanotechnology discourse and its elision of the word "nanotechnology" and the nano-prefixed technical lexicon helped to blur the scope of nanomaterial and nanomaterial-enabled goods manufacturing enterprises. For example, the President's Council of Advisors on Science and Technology (PCAST, 2010), citing Lux Research, reported in 2010 that some $224 billion of products making some use of "nanotechnology" components were sold worldwide in 2009. But the value of these components was relatively small, only $29 billion, $11 billion of which were manufactured in the United States (PCAST, 2010, p. 19). Many "nano" consumer products consisted of applications of nanoscale particles of prosaic materials such as silver, platinum, carbon, and titanium dioxide. These substances were distinct from truly novel materials like single-walled carbon nanotubes and quantum dots that were supposed to endow macroscale technologies with special properties.[12] The latter proved difficult to produce cheaply and apply and remained largely experimental as of 2010 (Bourzac, 2010; DOE, 2010). In the 2000s, many if not most makers of nanomaterials produced quantities far smaller than the 10,000-kilogram threshold requiring advance notice to the Environmental Protection Agency (EPA) (Davies, 2006). And existing aerosol mapping techniques worked best in manufactories employing continuous, not batch, production (NIOSH, 2009).

These should be considered factors in the inconclusive nanomaterial regulatory process, a debate prolonged by the inability of the various actors and agencies of regulatory science to agree on definitions and set boundaries (Klaessig, 2010). During the 2000s, the EPA did not regard size as a criterion in deciding chemical novelty under the Toxic Substances Control Act (TSCA). Its stance implied that existing rules encompassed nanomaterials, although by 2010 the agency was developing a Significant New Use Rule regulating production of new nanoscale materials based on substances on the TSCA list (EPA, 2010). In the early to mid-2000s, scientists raised the question of the broader public health implications of nanotechnology, justifying programs of early-stage risk assessment and education lest fearful citizens put up political roadblocks (Rip, 2006). But such concerns misconstrued actual public attitudes about nanotechnology at that time, which were linked primarily with economic and security issues (Mody, 2008). In 1999, the interagency group forecast that most jobs and profits in "nanotechnology" in the near term were likely to come from supplying costly specialized instruments and materials to the research and development apparatus (IWGN, 1999). If this prediction comes to pass, nanotechnology's material practices, and, hence, risks, will be confined to laboratory environments for years to come.

AN EVANESCENT DISCURSIVE FRAME?

A product of futurists, academic scientists, and policymakers staking a variety of claims in a period of political flux, the nanotechnology frame expressed real progress in existing fields of materials research through the traditional discourse of federal science policy, one that justified basic science as an economic engine. Part of what sustained nanotechnology rhetoric at its zenith in the early 1990s to the mid-2000s was its metaphorical versatility. It could be made to help justify a contradictory brew of ideas relating to innovation in the post–Cold War period (McCray 2005). Certain scientists believed its matter-as-machine metaphysics conferred political cover and prestige at a time when they were under pressure to demonstrate the practical value of their work. Technologies of quantum characterization and control enabled them to visually render the concept of "nanotechnology," making it intelligible to lay audiences and allowing them to take on the identities of professionals who manipulated matter dramatically (and often profitably) in the macroworld. For Smalley (2000), working with nanotechnology meant chemists were not just chemists but "builders" and "architects" as well. Marrying the concepts of radical physical and biological change with social stasis, nanotechnology corresponded well with the tenets of neoliberal capitalism. An allegory for market forces, it avoided the "centralized planning" of top-down fabrication (Sargent, 2006).

Yet nanotechnology illustrated idealized social relations, as Schummer (2010) has argued. This was not the first time physical scientists claimed expertise in the social relations of science and technology. In the postwar period, notes Mirowski (2002, 2004), they frequently sought to be consulted on how to scientifically manage the economy without actually becoming social scientists themselves. In promoting the folk sociology of nanotechnology, physical scientists and their representatives presented old arguments for the economic utility of science in a new form. Perhaps not surprisingly, they misread the nuances of language in Washington science politics. In late 1999, an official in the Office of Management and Budget warned Dehmer that the agency was more interested in supporting a science initiative than a technological one. He felt the NNI would run afoul of legislators in February and wished it could be renamed. In an e-mail to Smalley, Dehmer worried that the NSF Engineering Directorate's vanguard role in promoting the initiative would further complicate its legislation; the chemist responded that "only great scientists" would work in the nanotechnology field, citing the precedent of biotechnology (Dehmer, 1999). The result seems to have been a last-minute campaign by boosters to reassure federal bureaucrats of the NNI's scientific content.

Did the use of the nanotechnology frame in science policy discourse have broader social significance? Nanotechnology and biotechnology shared common origins in the speculative economy of collaborative research and development that emerged in the 1980s and 1990s. Michael Fortun (2001), speaking of genomics, characterized this economy as an attempt to write a "future already half-conjured by present statements" (p. 153). Merrill Lynch (2004), for example, selected the companies of its 2004 "Nanotech Index" partly on the basis of anticipated profits. But nanotechnology also became a buzzword of academic entrepreneurialism for managers and administrators who sensed more immediate opportunities in the get-rich-quick zeitgeist of the high-tech 1990s. In 1998, MIT shifted the mandate of its 99-year-old *Technology Review* from critical social analysis to the promotion of innovation, focusing on information technology, biotechnology, and nanotechnology; the short-term goal was to staunch annual losses of $200,000 by attracting new readership and advertising among the business class (Lohr, 1998).[13] Above all, nanotechnology rhetoric conveyed an ongoing commitment to the premise of federal science as a national economic engine, even as the tech bust of the early 2000s and the post-2008 recession indicated its role in this regard was relatively minor at best.

In this sense, nanotechnology represented continuity in science policy discourse. Accordingly, its social ambit was relatively circumscribed. Targeting fractious lawmakers who nevertheless declined to fundamentally alter the structure of the federal science establishment, nanotechnology rhetoric hardly impinged on the public imagination. Over half of the participants in one study conducted in the late 2000s evinced no knowledge of nanotechnology at all, with 30 percent reporting some knowledge (Satterfield, Kandlikar, Beaudrie, Conti, & Harthorn, 2009). Code for individualism and interdisciplinarity, for academic science liberated from the fusty Mertonian fetters of communalism and disinterestedness (public knowledge produced with no regard for utility), for current basic research applied in the deep future, the rhetoric of nanotechnology was a product of the short-lived crisis in the politics of science in the 1990s. With the finances of federal science as secure as ever in the 2000s, nanotechnology appeared to lose its *raison d'être*. Media attention in the United States peaked around 2006 and 2007, falling precipitously thereafter (Weaver, Lively, & Bimber, 2009; see also Chapter 11, this volume).

Mowery (2011) warned that the intensive patenting of nanomaterials might inhibit basic science if it restricted exchanges of materials. But nanotechnology's relatively low profile as a discursive frame and the public's lack of leverage over U.S. science policy suggest there may well be only limited fallout from the "novelty trap," the erosion of trust said to occur when the promotion of revolutionary new technologies is counterpoised by simultaneous appeals to continuity as a means of moderating expectations of risk (Rayner, 2004). However, the history of nanotechnology does illustrate the marginalization of social scientists in the process of shaping science policy discourse in the 1990s. They could play no substantive role in the project to develop nanotechnology as a discursive frame because it was the purview of a small elite group of physical scientists working largely behind closed doors. And although social scientists were enlisted by the federal government to study nanotechnology's ethical, legal, and social implications in the early 2000s, and later successfully deconstructed many folk theories of nanotechnology, they have not been in a position to challenge them at the level of science policymaking. Nevertheless, the story of nanotechnology as a social phenomenon (i.e., as discourse deployed to advance the traditional material interests of clients of the federal science establishment) has great potential value as a way of informing public awareness of the nature and scope of this set of institutional social relations and its ability to shape social landscapes.

ACKNOWLEDGMENTS

This material is based upon work supported by the National Science Foundation under Cooperative Agreement Nos. SES 0531184 and SES 0938099. Any opinions, findings, and conclusions, or recommendations expressed in this material are those of the author and do not necessarily reflect the views of the National Science Foundation. I am grateful to Rich Appelbaum, Gwen D'Arcangelis, Barbara Herr Harthorn, Mikael Johansson, Sharon Ku, W. Patrick McCray, Cyrus C. M. Mody, John W. Mohr, and Yasuyuki Motoyama for their constructive criticism during the drafting of this article. All errors are my own.

NOTES

1. McCray introduced the term "hidden histories" to describe this approach. Mody coined the similar expression "nano before there was nano" to title a conference held at the Chemical Heritage Foundation in March 2005 devoted to this subject (Chemical Heritage Foundation, 2005).
2. The career paths of Drexler and the futurists Aubrey de Grey, a computer scientist and "theoretical gerontologist," and Drew Endy, a civil engineer, are strikingly similar. Active in the 2000s and trained in disciplines unrelated to the life sciences, de Grey and Endy posited idealized "fields" of biochemical engineering, exploiting the notion of control connoted by the word "engineering" to impart certitude to a realm of research hitherto characterized by complexity (see Campos, 2009; Lafontaine, 2009).
3. George E. Brown (D-CA), ranking minority member of the House Science Committee and an ardent champion of science, supported the principle of cutting bureaucracy and "corporate welfare." Brown even claimed that federal agencies could be cut by up to one-third and still remain effective (1995, September 8, House hearing on the Department of Energy national laboratories. *FYI: The American Institute of Physics Bulletin of Science Policy News*, 124. Retrieved September 2011, from http://www.aip.org/fyi/1995/fyi95.124.htm). I have made PDF copies of all web pages cited in this article and will make them available on request.
4. NSF Director Lane: Understanding the issues and entering the fray. 1995, July 7. *FYI: The American Institute of Physics Bulletin of Science Policy News*, 96. Retrieved September 2010, from http://www.aip.org/fyi/1995/fyi95.096.htm.
5. NSF director Neal Lane on government shutdown. 1996, January 19. *FYI: The American Institute of Physics Bulletin of Science Policy News*, 8. Retrieved September 2010, from http://www.aip.org/fyi/1996/fyi96.008.htm.

6. Giant magnetoresistance-enabled technology (or "spintronics") was especially important to nanotechnology advocates as an example of actual "nanotechnology" produced through linear collaborative research and development (IWGN, 1999). However, this was less than an ideal case of national linear innovation, given that the basic research had been performed in Europe and applied and commercialized without state assistance by IBM, an established American company (see McCray, 2009).
7. President Clinton's address to Caltech on science and technology. 2000, January 21. Retrieved November 2010, from http://pr.caltech.edu/events/presidential_speech/
8. As enacted for fiscal year 2001, the CCTI received $1.239 billion for research and technology programs, whereas tax incentives were unfunded (see Simpson, 2001).
9. *Minutes for the Basic Energy Sciences Advisory Committee Meeting, November 3–4 1999*. Series III, Event Files, Box 33, Folder 9, Richard E. Smalley Papers. Philadelphia, PA: Chemical Heritage Foundation.
10. Of 63 journals identified under the subject category of "nanoscience and nanotechnology" by the ISI Web of Knowledge in 2011, only 11 contained "nanotechnology" somewhere in their title. These are, with their founding dates: *IEEE Transactions on Nanotechnology* (2002), *IET Nanobiotechnology* (2007), *International Journal of Nanotechnology* (2004), *Journal of Biomedical Nanotechnology* (2005), *Journal of Nanoscience and Nanotechnology* (2001), *Nanomedicine: Nanotechnology Biology and Medicine* (2005), *Nanotechnology* (1990), *Nature Nanotechnology* (2006), *Precision Engineering-Journal of the International Societies for Precision Engineering and Nanotechnology* (1979), *Recent Patents on Nanotechnology* (2007), and *Wiley Interdisciplinary Reviews-Nanomedicine and Nanobiotechnology* (2009). See http://admin-apps.webofknowledge.com/JCR/JCR?RQ=LIST_SUMMARY_JOURNAL&cursor=1 retrieved September 2011.
11. In 2011, the NNI homepage listed 11 bachelor's, 10 master's, and some 59 doctoral degree programs in nano-prefixed fields, almost all in the United States. The list is somewhat misleading, as virtually none of the cited degrees were in "nanotechnology," as the NNI website acknowledged in a disclaimer. Seven or eight institutions offered a master's of science or engineering degree in a nano-prefixed field such as nano-engineering or nanoscience, with nanotechnology being the least common descriptor. All of these programs were typically housed in traditional physics, chemistry, chemical engineering, engineering, or materials sciences departments. Moreover, only a handful of institutions offered a PhD in these fields (University of North Carolina at Charlotte, the University of Albany's College of Nanoscale Science and Engineering, and the South Dakota School of Mines and Technology); See "University Education," retrieved April 2011, from http://www.nano.gov/html/edu/eduunder.html.
12. Multi-walled carbon nanotubes (MWCN) are structurally different than single-walled carbon nanotubes (SWCN), so much so that they deserve to be classified separately. Nevertheless, they are sometimes lumped together (e.g., see Lux Research, 2007). Relatively cheap ($.10 per gram), MWCNs have structural defects that render them unsuitable as a material for the atomically perfect nanostructures that Smalley and others envisioned as the basis for revolutionary electronics and power equipment. For such roles, SWCN are prized. But these have proved much more difficult and costly ($200–$900 per *milligram*) to produce in volume. In the early 2000s, most manufacturers were academic institutions preparing small amounts for research purposes (see Ouellette, 2002/2003; retrieved March 2010, from http://www.nanointegris.com/en/metallic-m).

13. The new direction of *Technology Review* did not completely silence skeptical voices. For example, the January/February 2000 issue carried a scathing review of molecular computing by Zachary (2000), who referred to it as the "Big Lie."

REFERENCES

Auffan, M., Rose, J., Bottero, J., Lowry, G. V., Jolivet, J., & Wiesner, M. R. (2009). Towards a definition of inorganic nanoparticles from an environmental, health, and safety perspective. *Nature Nanotechnology, 4*, 634–641. Published online: September 13, 2009 | DOI:10.1038/nnano.2009.242

Bazerman, C. (1988). *Shaping written knowledge: The genre and activity of the experimental article in science.* Madison: The University of Wisconsin Press.

Bensaude-Vincent, B. (2004). Two cultures of nanotechnology? *HYLE-International Journal for Philosophy of Chemistry, 10*(2), 65–82.

Bimber, B. (1996). *The politics of expertise in Congress: The rise and fall of the Office of Technology Assessment.* Albany: State University of New York Press.

Block, F. (2008). Swimming against the current: The rise of a hidden developmental state in the United States. *Politics and Society, 36*(2), 169–206.

Bourzac, K. (2010). Nanotube fibers: Solutions of carbon nanotubes can be used to make strong, conductive fibers hundreds of meters long. *Technology Review, 113*(3), 92–94.

Campos, L. (2009). That was the synthetic biology that was. In M. Schmidt, A. Kelle, A Ganguli-Mitra, H. de Vriend, et al. (Eds.), *Synthetic biology: The technoscience and its societal consequences*, pp. 5–22. Dordrecht: Springer.

Chemical Heritage Foundation. (2005, March 19). Nano before there was nano: Historical perspectives on the constituent communities of nanotechnology. Cain Conference. Retrieved August 2011, from http://h-net.msu.edu/cgi-bin/logbrowse.pl?trx=vx&list=h-sci-med-tech&month=0409&week=d&msg=hFNryYz1f5X9UIu2TfFiig&user=&pw

Choi, H., & Mody, C. C. M. (2009). The long history of molecular electronics: Microelectronics origins of nanotechnology. *Social Studies of Science, 39*(1), 11–50.

Clinton, W. J. (2000, January 21). President Clinton's address to Caltech on science and technology. Retrieved November 2010, from http://pr.caltech.edu/events/presidential_speech/

Clinton, W. J., & Gore, A. Jr. (1993, February 22). Technology for America's economic growth: A new direction to build economic strength. Retrieved November 2010, from ntl.bts.gov/lib/jpodocs/briefing/7423.pdf.

Davies, J. C. (2006). *Managing the effects of nanotechnology.* Washington, DC: Woodrow Wilson International Center for Scholars, Project on Emerging Nanotechnologies.

Dehmer, P. (1999). Dehmer to Smalley, November 6, 1999; Smalley to Dehmer, November 8, 1999. Series VI, Subject Files, Box 57, Folder 8, Richard E. Smalley Papers. Philadelphia, PA: Chemical Heritage Foundation.

Drexler, K. E. (1986). *Engines of creation: The coming era of nanotechnology.* Garden City, NY: Anchor Press/Doubleday.

Forman, P. (2007). The primacy of science in modernity, of technology in postmodernity, and of ideology in the history of technology. *History and Technology, 23*(1–2), 1–152.

Fortun, M. (2001). Mediated speculations in the genomics futures markets. *New Genetics and Society, 20*(2), 139–156.

Gallo, J. (2009). The discursive and operational foundations of the national nanotechnology initiative in the history of the National Science Foundation. *Perspectives on Science, 17*(2), 174–211.

Greenberg, D. S. (2001). *Science, money, and politics: Political triumph and ethical erosion*. Chicago: University of Chicago Press.
Guston, D. H. (2000). *Between politics and science: Assuring the integrity and productivity of research*. Cambridge: Cambridge University Press.
Johnson, A. (2004). The end of pure science: Science policy from Bayh-Dole to the NNI. In D. Baird, A. Nordmann, & J. Schummer (Eds.), *Discovering the nanoscale*, pp. 217–230. Amsterdam: IOS.
Johnson, A. (2009). Modeling molecules: Computational nanotechnology as a knowledge community. *Perspectives on Science, 17*(2), 144–173.
Harris, R. A. (1997). Introduction. In R. A. Harris (Ed.), *Landmark essays on rhetoric of science: Case studies*, pp. xi-xlv. Mahwah, NJ: Hermagoras Press.
Interagency Working Group on Nanoscience, Engineering and Technology (IWGN). (1999). *Nanotechnology research directions: IWGN workshop report vision for nanotechnology R&D in the next decade*. Baltimore, MD: International Technology Research Institute, World Technology (WTEC) Division, Loyola College.
Kim, E. (2008). Directed evolution: A historical exploration into an evolutionary experimental system of nanobiotechnology, 1965–2006. *Minerva, 46*(4), 463–484.
Kleinman, D. L. (1995). *Politics on the endless frontier: Postwar research policy in the United States*. Durham, NC: Duke University Press.
Lafontaine, C. (2009). Regenerative medicine's immortal body: From the fight against ageing to the extension of longevity. *Body and Society, 15*(4), 53–71.
Lane, N. (2001). The grand challenges of nanotechnology. *Journal of Nanoparticle Research, 3*(2–3), 95–103.
Lane, N., & Kalil, T. (2005). The National Nanotechnology Initiative: Present at the creation. *Issues in Science and Technology, Summer*, 49–55.
Lohr, S. (1998, April 20). MIT re-engineers magazine to attract new readers and ads. *New York Times*, p. D7.
Lux Research. (2007). *The nanotech report: Investment overview and market research for nanotechnology* (5th ed.). New York: Author.
Markoff, J. (1999, November 1). Computer scientists are poised for revolution on tiny scale. *New York Times*, p. C1.
McCray, W. P. (2005). Will small be beautiful? Making policies for our nanotech future. *History and Technology, 21*(2), 177–203.
McCray, W. P. (2009, January). From lab to iPod: A story of discovery and commercialization in the post-Cold War era. *Technology and Culture, 50*(1), 58–81.
Merrill L. (2004, April 8). Nanotechnology: Introducing the Merrill Lynch Nanotech Index. Retrieved November 2010, from http://www.ml.com/media/42322.pdf
Milburn, C. (2008). *Nanovision: Engineering the future*. Durham, NC: Duke University Press.
Mirowski, P. (2002). *Machine Dreams: Economics becomes a cyborg science*. New York: Cambridge University Press.
Mirowski, P. (2004). *The effortless economy of science?* Durham, NC: Duke University Press.
Mirowski, P. (2011). *Science-mart: Privatizing American science*. Cambridge, MA: Harvard University Press.
Mody, C. C. M. (2004). How probe microscopists became nanotechnologists. In D. Baird, A. Nordmann & J. Schummer (Eds.), *Discovering the nanoscale*, pp. 119–133. Amsterdam: IOS Press.
Mody, C. C. M. (2006). Corporations, universities, and instrumental communities: Commercializing probe microscopy, 1981–1996. *Technology and Culture, 47*(1), 56–80.

Mody, C. C. M. (2008). The larger world of nano. *Physics Today, 61*(2), 38–44.
Mody, C. C. M. (2009). Introduction. *Perspectives on Science, 17*(2), 111–122.
Mody, C. C. M. (2010). *Institutions as stepping stones: Rick Smalley and the commercialization of nanotubes.* Philadelphia, PA: Chemical Heritage Foundation.
Motoyama, Y., Appelbaum, R., & Parker, R. (2011). The National Nanotechnology Initiative: Federal support for science and technology, or hidden industrial policy? *Technology in Society,* 33(1–2), 109–118.
Mowery, D. C. (2011). Nanotechnology and the U.S. national innovation system: Continuity and change. *Journal of Technology Transfer, 36*(6), 697–711. DOI 10.1007/s10961–011–9210–2
National Research Council. (2009). *Review of the federal strategy for nanotechnology-related environmental, health, and safety research.* Washington, DC: The National Academies Press.
Nordmann, A. (2009). Invisible origins of nanotechnology: Herbert Gleiter, materials science, and questions of prestige. *Perspectives on Science, 17*(2), 123–143.
Ouellette, J. (2002/2003, December/January). Building the future with carbon tubes. *Industrial Physicist: American Institute of Physics, 8*(6), 18–21.
President's Council of Advisors on Science and Technology (PCAST) (2010, March 12). *Report to the President and Congress on the Third Assessment of the National Nanotechnology Initiative.* Retrieved August 2010, from www.whitehouse.gov/sites/default/files/. . ./pcast-nano-report.pdf
Rayner, S. (2004). The novelty trap: Why does institutional learning about new technologies seem so difficult? *Industry and Higher Education, 18*(6), 349–355.
Reagan, R. (1985, February 27). Remarks at the presentation ceremony for the National Medal of Science. Retrieved September 2011, from www.reagan.utexas.edu/archives/speeches/1985/22785a.htm
Rip, A. (2006). Folk theories of nanotechnologists. *Science as Culture, 15*(4), 349–365.
Rosenberg, N. (1994). *Exploring the black box: Technology, economics, and history.* New York: Cambridge University Press.
Sargent, T. (2006). *The dance of molecules: How nanotechnology is changing our lives.* New York: Thunder's Mouth Press.
Satterfield, T., Kandlikar, M., Beaudrie, C. E. H., Conti, J., & Harthorn, B. H. (2009). Anticipating the perceived risk of nanotechnologies. *Nature Nanotechnology, 4*, 752–758. DOI: 10.1038/NNANO.2009.265
Schummer, J. (2006). Gestalt switch in molecular image perception: The aesthetic origin of molecular nanotechnology in supramolecular chemistry. *Foundations of Chemistry, 8*(1), 53–72.
Schummer, J. (2010). From nano-convergence to NBIC-convergence: "The best way to predict the future is to create it." In M. Kaiser, M. Kurath, S. Maasen, C. Rehmann-Sutter (Eds.), *Governing future technologies: Nanotechnology and the rise of an assessment regime*, pp. 57–71. Dordrecht: Springer.
Siegel, R. W., Hu, E., Cox, D. M., Goronkin, H., Jelinski, L., Koch, C. C., Mendel, J. Roco, M.C., Shaw, D.T. (1999). *Nanostructure science and technology: A worldwide study.* Baltimore, MD: Loyola College, International Technology Research Institute, World Technology Division. Retrieved November 2010, from http://itri.loyola.edu/nano/final/
Simpson, M. M. (2001, October 3). Climate change: Federal research, technology, and related programs. *Congressional Research Service Report RL30452*, pp. 1–13. Retrieved February 2011, from www.nationalaglawcenter.org/assets/crs/RL30452.pdf

Smalley, R. E. (1993). Smalley to Carroll and Kinsey, January 21, 1993. Series II, Correspondence, 1992–1996, Box 3, Folder 3, Richard E. Smalley Papers. Woodson Research Center, Fondren Library, Rice University, Houston, Texas.

Smalley, R. E. (2000, January 22). Bucky balls, fullerenes, and the future: An oral history with Professor Richard E. Smalley. Interview by Robbie Davis-Floyd and Kenneth J. Cox. Rice University.

Smith, B. L. R. (1990). *American science policy since World War II*. Washington, DC: The Brookings Institution.

Taniguchi, N. (1974). On the basic concept of "Nano-Technology." *Proceedings of the International Conference on Production Engineering, Tokyo 1974 (Part II)*, pp. 18–23. Tokyo: The Japan Society of Precision Engineering.

U.S. Department of Energy (DOE). *Office of energy efficiency and renewable energy, industrial technologies program, nanomanufacturing, R&D portfolio*. Retrieved November 2010, from http://www1.eere.energy.gov/industry/nanomanufacturing/portfolio.html

U.S. Environmental Protection Agency (EPA). *Control of nanoscale materials under the Toxic Substances Control Act*. Retrieved February 2010, from http://www.epa.gov/oppt/nano/#pmn

U.S. House of Representatives. (1999, June 22). *Nanotechnology: The state of nano-science and its prospects for the next decade* (Publication No. 106–40). Washington, DC: US Government Printing Office. Retrieved August 2010, from www.gpo.gov/fdsys/pkg/CHRG. . ./html/CHRG-106hhrg60678.htm

U.S. National Institute for Occupational Safety and Health. (2009, March). *Approaches to safe nanotechnology: Managing the health and safety concerns associated with engineered nanomaterials*. Centers for Disease Control and Prevention. Retrieved February 2010, from http://www.cdc.gov/niosh/docs/2009-125/

U.S. National Nanotechnology Initiative (NNI). (2011). *University education*. Retrieved August 2011, from http://www.nano.gov/html/edu/eduunder.html

U.S. National Science Board (NSB). (1996). *Science and engineering indicators 1996* (NSB 96–21). Washington, DC: U.S. Government Printing Office.

U.S. National Science Board (NSB). (2010). *Science and engineering indicators 2010* (NSB 10–01). Arlington, VA: National Science Foundation.

Weaver, D. A., Lively, E., & Bimber, B. (2009). Searching for a frame: News media tell the story of technological progress, risk, and regulation. *Science Communication, 31*(2), 139–166.

Zachary, G. P. (2000). Nano-hype: If you think a self-assembling molecular computer is right around the corner, you may have succumbed to the big lie. *Technology Review, 103*(1), 39.

Ziman, J. (2000). *Real science: What it is, and what it means*. Cambridge: Cambridge University Press.

3 When Space Travel and Nanotechnology Met at the Fountains of Paradise

W. Patrick McCray

In recent years, there has been much debate among pundits and prognosticators about "converging technologies" or "technological singularities." Simply put, these visions hold that sometime in the twenty-first century human society will arrive at a point when rapidly accelerating technological change will bring changes such as new industrial revolutions, unity between fields of scientific research, and perhaps radical changes in human intelligence and capabilities (Roco & Bainbridge, 2003; Vinge, 1983).[1] Advocates of these "convergence" scenarios say it is not a matter of whether these technologies will come together, but simply when and how (Kurzweil, 2005). Critics, meanwhile, say such predictions are rooted in the selective observations about previous technological trends, with the singularity itself representing a "set of untestable assumptions about our near future" (Hassler, 2008, p. 9).

Utopian beliefs about the transformative effects of technology are not new, of course. They typically reflect a progressive vision of the future rooted in a technologically determinist framework (McCray, 2005; Segal, 1985). In the mid-twentieth century, *the* technological frontier was, of course, space exploration. Scholars, scientists, and journalists predicted that an expanded human presence in space was inevitable with profound societal and economic changes to follow (Mazlish, 1965; McCurdy, 1997). The first decades of human space exploration generated huge amounts of media attention, inspired schoolchildren to pursue careers in science, and motivated governments to spend billions of dollars in competition with one another. The similarities between imaginations of the space future in the 1960s resonate strongly with those for nanotechnology and singularity-like scenarios circa 2001.

Throughout history, of course, technologies have converged in many creative and surprising ways (Thompson, 2009). In the 1950s and 1960s, for example, the tools for space exploration were the example *par excellence* of a converging technology. Engineering studies and fictional scenarios for space exploration brought together frontier computing technologies and studies of human physiology in extreme environments with advanced microelectronics and the development of new high-tech materials. These were all combined with the latest in aerospace and rocket technologies.

Over time, however, the public's attention and awareness of over-the-horizon technologies shifted as predictions of new technological convergences embraced new frontiers. These included biotechnology, which emerged in the 1970s, cyberspace and the Internet in the late 1980s, and, more recently, nanotechnology which futurists and government officials held out as *the* preeminent technology of the twenty-first century.[2]

Over the past 40 years, space and nanotechnology have crossed paths, linked, and gone their separate ways several times (so often, in fact, that we might consider them "reconverging technologies"). At times, connections between the two technologies have been in the realm of engineering design, and at other times, they have resided in science fiction and futuristic scenarios. To help illustrate the ways in which aspirations for both the conquest of outer space and nano-space have converged in detailed engineering studies, laboratory research, and science fiction scenarios, this chapter takes a look at the designs and plans for a hypothetical technological artifact: the space elevator.

The basic idea of a space elevator is straightforward. A tether or cable, made of a super-strong and lightweight material, would be anchored somewhere in the equatorial regions to a platform with facilities similar to a modern airport or rocket launch site. The cable would reach upwards from this base thousands of miles into geostationary orbit. The other end of the cable might be connected to a counterweight in space, perhaps a small asteroid. As the earth rotates, the inertia of the cable's end keeps it taut and counteracts gravitational pull. Meanwhile, climbers (analogous to elevator cars) powered by solar cells or laser propulsion could ascend the cable and move people and cargo into orbit. Advocates of the concept argue that, once built, the space elevator could provide a cheaper and more environmentally friendly alternative to blasting payloads into orbit via nonreusable rockets. Skeptics say that although such a technology breaks no laws of physics—this is not the realm of warp drives or antigravity machines—a space elevator cannot be built with currently available materials, and even in the future, the political and economic challenges would be considerable (Chang, 2003).

The idea that a cable or tower anchored to earth and stretching out into space could put people and payloads into orbit is more than a century old. Despite numerous thought experiments and technical studies, the feasibility of actually building a space elevator remained purely speculative until the 1990s. The primary obstacle was the lack of any viable material to fashion a cable strong yet light enough to serve as the elevator's backbone. However, with the discovery of carbon nanotubes in the 1990s, advocates for a space elevator claimed that nanotechnology might offer a possible solution to the designer's dilemma, perhaps even a path to the stars. As a result, the space elevator, even though it remains a hypothetical technological artifact, provides a bridge between two vastly different scales of techno-scientific exploration, outer space and nanotechnology.

MERGING OUTER SPACE WITH NANOSPACE

For more than three decades, technological enthusiasts have linked a fascination with space exploration to a whole host of future-looking technologies, including nanotechnology, artificial intelligence, and robotics. The word "cyborg," for example, has its origins in the work of Manfred Clynes and Nathan Kline, two researchers interested in altering the human body via drugs and biological modification in order to allow astronauts to survive the extreme environs of outer space (Kline, 2009; Kline & Clynes, 1961). Proposed even before Yuri Gagarin orbited the earth, Clynes' and Kline's "cybernetic organisms" presaged scenarios for converging technologies that advocates of transhumanism described years later (Bostrom, 2005).

The links between nanotechnology and space exploration go back, not surprisingly, to the activities and writing of K. Eric Drexler, the most visible popularizer of nanotechnology starting in the mid-1980s and continuing well into the 1990s. Before this, however, Drexler was a devoted supporter of space-based settlements and manufacturing, ideas promulgated largely through the technical studies and writings of Princeton physicist Gerard O'Neill. Drexler first encountered O'Neill's ideas about the "humanization of space" as an undergraduate student at MIT in the 1970s. As an active member of the L5 Society, a citizens' pro-space group from that era, Drexler wrote technically sound articles about the future possibilities of asteroid mining, space-based closed ecological systems, and solar sails (even receiving patents for his ideas in the process).

After moving to Silicon Valley in 1985, Drexler published a book that summarized his vision of the technological future. Titled *Engines of Creation*, Drexler's book was written for broad, nonspecialized readers interested in the future of technology and its implications. With an introduction by Marvin Minsky, MIT's guru of artificial intelligence and one of Drexler's mentors, *Engines* became the canonical text for the specific form of nanotechnology Drexler popularized (Drexler, 1986).[3] A major theme of *Engines*, one promoted by other early pro-nanotech enthusiasts of whom many had been involved in the pro-space movement, was that nanotechnology offered solutions to a whole host of problems, including the need for cheaper and easier ways to access space.

In the mid-1990s, one of the first government agencies in the United States to support an in-house effort to explore the possibilities of nanotechnology was the National Aeronautics and Space Administration (NASA). At its Ames Research Center, located in the heart of Silicon Valley (where futuristic technologies of all sorts are the coin of the realm), a small group of chemists and computer experts started a research program in computational nanotechnology. Besides its access to Silicon Valley's computer communities, Ames hosted NASA's supercomputing center. Originally developed to do simulations of airflow and other fluid dynamics problems, the center's high-speed parallel computing resources could also be directed to molecular nanotechnology simulations.

The group's initial goal, according to an early member, was to explore some of the more radical Drexlerian ideas for nanotechnology and see how they might be applied to space-related technologies (Johnson, 2006; A. Globus, personal communication, September 24, 2009). The new forms of carbon discovered recently, buckyballs (or fullerenes) and nanotubes, received a great deal of the group's attention. They discussed the possibility that "in the distant future," one could design and build "atomically precise programmable machines" made of fullerenes. To back this up, the group did extensive design and simulations of nanoscale gears, sensors, and other devices "built and tested" via computer. Some 65 people were part of the Ames nanotechnology team at its peak, making it one of the world's largest groups focused on nanoscale research. But, as the group continued to expand, its research shifted away from computer modeling of nanoscale gears and machines with Ames's supercomputers to the lab-based fabrication and study of actual materials and devices, including, eventually, carbon nanotubes.

NASA's initial work reflected what Drexler called "exploratory engineering." By this, he meant the practice of "designing things we can't yet build" (1988, p. 132); although perhaps anathema to some scientists, engineers have often designed objects which are possible but not within the reach of current manufacturing. Although their designs were sound, they called for new materials and other technologies that, although imaginable, did not exist yet. It is often through this process that engineers develop a better understanding of the weaknesses of current designs and work out paths to achieve improvements. Indeed, all engineering is heavily oriented toward the future—asking about what could be built, designing plans for what might be built (Constant, 1980; Johnson, 2010; Vincenti, 1990).

Even outside the world of futuristic technology prognosticators and small groups of NASA researchers, links—rhetorically, at least—between space and nanotech existed. In 1999, policymakers in the U.S. government were laying the groundwork for a major new research initiative that promoted nanotechnology as the next technological revolution. In advance of announcing the National Nanotechnology Initiative (NNI), advocates produced a glossy brochure that highlighted the research bonanza that nanotechnology promised. The brochure's cover juxtaposed an image of a crystalline surface made with a scanning tunneling microscope, nanotech's counterpart to the rocket in terms of iconic exploratory technology, with "cosmic imagery" (Nordmann, 2004) that "evokes the vastness of nanoscience's potential" (Interagency Working Group, 1999). That same year, as scientists and engineers lobbied Congress for the NNI, Nobel laureate

Richard E. Smalley invoked the spirit of the Apollo program. What was needed, Smalley said, was someone bold enough to "put a flag in the ground and say: 'Nanotechnology, this is where we are going to go'" (House of Representatives Committee on Science, 1999, p. 12).

Rhetorical flourishes and symbolism aside, when federal science managers organized the NNI they ignored its direct historical ties to futurism and exploratory engineering—what my colleagues and I have termed hidden or camouflaged history—and focused instead on more immediate needs, for instance, of the electronics industry and other sectors connected to economic competitiveness (McCray, 2005). Whether motivated by embarrassment or expediency, the futuristic aspects of nanotechnology's history were detached, at least officially, from federal policy initiatives. An examination of the space elevator's history helps us reconnect some of these severed threads.

THE SPACE ELEVATOR, INVENTED OVER AND OVER

The Russian space visionary Konstantin Tsiolkovsky once wrote, "First, inevitably, the idea, the fantasy, the fairy tale. Then, scientific calculation. Ultimately, fulfillment crowns the dream" (Crouch, 1999, p. 30). This path from imagination to reality is a condensed history of a whole host of radical technological ideas that have yet to be realized, including the space elevator.

The idea that one could anchor a cable, stretch it out into space, and use it to transport people and payloads in and out of earth's gravity well is more than a century old. In 1895, Tsiolkovsky published a series of thought experiments that elevator advocates cite as a starting point for later detailed studies (Rynin, 1971).[4] One idea Tsiolkovsky proposed was a hypothetical tower extending into space, a place he used to imagine a gravity-free environment (Pearson, 1997).[5] Tsiolkovsky soon turned his attention to developing designs and theories for rocket motors—exploratory engineering that space pioneers like Robert Goddard, Frank Malina, and Wernher von Braun later proved right—and his earlier idea of a tower reaching to the heavens was mostly forgotten until the 1960s.

During the Space Race, Soviet and American scientists and engineers proposed general concepts, incorporating space-to-earth cables several times in the technical literature (Artsutanov, 1960; Lvov, 1967; Pearson, 1975). However, rockets, not "heavenly funiculars," as one Russian writer called them, were the ascendant technology during the Apollo era. All other options remained engineering curiosities. For example, Isaacs, Vine, Bradner, and Bachus (1966) described a space-elevator-like concept in 1966, 3 years before Neil Armstrong and Buzz Aldrin walked on the moon. However, the authors of this peer-reviewed article in *Science* were researchers at oceanographic research laboratories, not NASA. Hans Moravec, who became a robotics guru and major proponent of transhumanist technologies, proposed a similar idea for a "sky hook" in the 1970s while he was studying artificial intelligence at Stanford (Moravec, 1977). The point here is that many of people promoting alternatives to the traditional rockets were relatively obscure in the broader space community or situated outside of it entirely.[6]

In 1979, science fiction author and futurist Arthur C. Clarke published his novel *The Fountains of Paradise* and thus ended the space elevator's relative invisibility. The plot of *Fountains* revolves around a visionary engineer (Vannevar Morgan, a certain nod to Vannevar Bush, the doyen of U.S. science policy), who designs a bridge to space that is located on a fictional version of Sri Lanka. In Clarke's story, engineers built the elevator's cable from "pseudo-one-dimensional diamond crystals" he called "hyperfilament" (p. 39). *Fountains of Paradise* introduced the space elevator to thousands of sci-fi readers, and Clarke promoted the idea via interviews and appearances at technical meetings. Clarke's enthusiasm helped establish credibility for the idea. He had, after all, predicted telecommunications satellites long before *Sputnik*, co-authored the screenplay for *2001:A Space Odyssey*, and sat next to Walter Cronkite during the first Apollo landings.

Although Clarke's ideas were firmly grounded in scientific and engineering fact, at least one major obstacle remained. As one space elevator advocate recalled, all of these early designs assumed a large supply of "unobtainium" (B. C. Edwards, personal communication, June 4, 2007).[7] The lack of a viable material from which to make a cable presented what engineers sometimes call a "tall tent pole"—a technical problem so critical that solving it can hold up the entire project, with all other problems lost beneath the canvas. The people who speculated on how to build a space elevator all agreed that the cable would have to be extraordinarily light and strong, and various studies considered Kevlar and materials containing graphite whiskers. Diamond, a material whose artificial manufacture received a great deal of attention from materials scientists in the 1980s, was also a candidate. All of these high-tech materials, however, were dismissed by space elevator advocates as impractical or impossible. Without a real material—one that actually existed and which might be manufactured in large amounts—further design work on a space elevator was beside the point.

SMALLEY'S SPACE

This situation changed after November 1991 when Sumio Iijima, a Japanese physicist, announced that he had made "helical microtubules of graphitic carbon." These long, thin cylinders, later christened carbon nanotubes, became another key icon of nanoscience (Iijima, 1991; Kroto, Heath, O'Brien, Curl, & Smalley, 1985). The ensuing flood of research on carbon nanotubes moved the concept of the space elevator into the realm of the theoretically possible. It also brought new players and institutions to the space–nanotechnology nexus.

Rice University professor Richard Smalley was one of these new arrivals. In 1985, *Nature* carried an article co-authored by Smalley, Curl, and Kroto that announced the discovery of carbon-60, christened "Buckminsterfullerene." This discovery set off an avalanche of research and publications on this new allotrope of carbon. Iijima's 1991 discovery was one of the outcomes of this work, and as Mody (2010) has shown, it catalyzed Smalley's interest in both carbon nanotubes as well as nanotechnology.

Smalley was willing to use Drexler's popular writings as a vehicle to raise research funds for his own work at Rice University. For example, in 1993, when Smalley wanted to generate support for a new nano-research center at Rice University, he mailed copies of Drexler's books to the school's trustees (Smalley, 1993a, 1993b). At the same time, Smalley had extensive correspondence with people about how NASA might support his research on carbon nanotubes. As he put it in an e-mail message to a Rice colleague, "I have some thoughts that will seem crazy at first blush . . . I'm talking with some people at NASA JSC [Johnson Space Center, located near the Rice campus] next week to see what they think about all this fantasizing. Maybe we can suck off a bit of their >$10,000,000,000/yr budget and actually do something worthwhile" (Smalley, 1996).

Smalley's main interest, of course, was to secure stable funding. This would enable his growing research network to better understand carbon nanotubes and eventually produce them in large amounts. Nonetheless, space also beckoned to Smalley who, as he stated in interviews, chose his career because of the Space Race. In public talks and private meetings with potential patrons, Smalley placed the space elevator between fact and fiction, suggesting that with the right support and funding it might be possible. If nothing else, it could serve to stimulate nanoscale research on more near-term goals and bring attention to the burgeoning new field.

For example, the colorful cover of the July–August 1997 issue of *American Scientist* depicted a nanotube-enabled elevator extending out into space. The feature article inside, coauthored by Smalley, detailed the science behind these new allotropes of carbon (Yakobson & Smalley, 1997). But when he wanted to illustrate the extraordinary theoretical strength of carbon nanotubes, Smalley, who was initially supportive of Drexler's visions for nanotechnology, turned to the possibility of using carbon nanotubes for a future space elevator. Smalley's overall strategy of blending futuristic and technologically revolutionary possibilities with near-term "normal science" worked. By 1999, NASA supported his Carbon Nanotechnology Laboratory at Rice with a multimillion-dollar award.

The road between science fact and fiction went both ways after the discovery of carbon nanotubes. Kim Stanley Robinson's 1993 novel *Red Mars* placed a space elevator at the center of a conflict between groups with competing goals for settling the Red Planet. Robinson's book, which won the Nebula Award in 1993, described a cable made from robotically processed carbon mined from asteroids. What form it is in is not clear—later in the book, he refers to a cable made from "graphite whisker with diamond sponge-mesh gel double-helixed into it" (p. 499). But, in homage to earlier sci-fi writers who used the elevator concept, Robinson named the two stations for the transport system "Clarke" and "Sheffield."[8]

The interest of Nobel laureate Smalley and sci-fi writers like Robinson shows where space exploration and nanotechnology converged again in the guise of a hypothetical space elevator. More importantly, it points to the willingness of "mainstream" scientists to embrace exploratory engineering in their approaches to the public and patrons. By the late 1990s, university scientists, business leaders, and policymakers had marginalized Drexlerian ideas, if not Drexler himself. Nonetheless, Smalley adopted the same rhetorical strategy Drexler often used of blending science fiction scenarios with scientific descriptions and engineering projections. By this point in the space elevator's history, the vision, and possibly the materials for fulfilling it, were imagined to be there. The next stage was to carry out more robust and thorough design studies.

"ONCE YOU HAD CARBON NANOTUBES . . ."

At about the time that Smalley's article in *American Scientist* appeared (Yakobson & Smalley, 1997), Bradley C. Edwards, a scientist at Los Alamos National Laboratory, was beginning to mull over the space elevator idea. Edwards, a physicist who helped build hardware for space missions, read that a space elevator could not be built for at least 300 years. The outright dismissal of the concept bothered Edwards, and he began to seriously explore the concept. As he explained, "Once you had carbon nanotubes, the rest of it could be done" (B. C. Edwards, personal communication, June 4, 2007). As he studied the existing technical literature, Edwards also read science fiction treatments by Clarke, Kim Stanley Robinson, and other authors. Edwards recalled, "While you can't actually build the cable that is in *Red Mars*, Robinson did enough careful thought and planning on it so that it wasn't too far off base. That was very interesting to me" (B. C. Edwards, personal communication, June 4, 2007).

For decades, even as the bulk of NASA's budget went to more traditional rocket and space science programs, the space agency maintained a modicum of support for advanced concepts so long as they were sufficiently far in the future to not endanger existing programs. One such operation was the NASA Institute for Advanced Concepts (NIAC). Created in 1998, the NIAC—a small operation that had an annual budget around $4.5 million—sought to "inspire, select, and fund revolutionary ideas and concepts" (R. Cassnova, personal communication, June 29,

2007). In 1999, Edwards requested funding from NASA to explore the idea further. After passing peer review, Edwards received some $570,000 over the next 3 years from NIAC and raised several million more from other sources (NIAC, 2007).[9] Between 2000 and 2003, Edwards thoroughly analyzed the entire space elevator idea, including safety factors, economic implications, and legal issues (see Edwards, 2003). His previous experience at Los Alamos designing spacecraft systems was an asset, and he divided his space elevator design into dozens of separate yet manageable technical problems. Whereas, for example, previous concepts for a space elevator imagined thick cables weighing hundreds of thousands of tons, the potential of carbon nanotubes allowed Edwards to design his hypothetical system around an "initial ribbon, 8 inches wide and thinner than paper" and based, of course, on carbon nanotubes (Edwards & Westling, 2003, p. 43).

Edwards' research reinvigorated the space elevator concept and helped generate a flood of media attention and public interest. Edwards, who retired from Los Alamos to pursue the idea full time, gave scores of invited talks, including audiences at the Air Force, DARPA, and Congress. Meanwhile, articles in major newspapers started to carry stories about the space elevator. With predictable titles like "Stairway to Heaven," these helped create a flood of interest. Web-based magazines and blogs stimulated more interest. To give one example of the situating of the space elevator between fiction and reality, the PBS science show *Nova* ran a segment on the space elevator. As the show's host, astrophysicist Neil deGrasse Tyson, said, "Fueled by the promise of these tiny tubes, people are already working to turn the Space Elevator into a reality . . . Perhaps someday technology will catch up with our imaginations and take the Space Elevator out of the realm of science fiction once and for all" (McMaster, 2007).[10] Humorist Dave Barry (2003) was less respectful: Government scientists, Barry wrote, had an idea "so radical . . . you wonder if they've been smoking reefers the size of Yule logs."

New groups, motivated by Edwards' exploratory engineering, began to get in on the action. In 2005, contributions from the Spaceward Foundation, a nonprofit organization "dedicated to furthering space science and technology," helped NASA add space-elevator related contests to its Centennial Challenges program. NASA's initiative, sought to encourage innovation by awarding cash prizes to teams meeting specific technical challenges.[11] These contests stimulated the space elevator community and broadened its base of support. For several years, scores of garage tinkerers and students met to compete for large prizes in contests to develop hardware for a future space elevator. In 2009, the *New York Times*, CNN, and dozens of web-based news sources reported on the Space Elevator Games (Chang, 2009). Held in Mojave, California, where several private spaceflight startups are based, the Games featured LaserMotive, a small Seattle company, which won $900,000 for demonstrating a prototype of a machine that might one day scale a space elevator's cable. The space elevator, at least as an incentive for innovation and experimentation, had left the ground floor.

SPECULATION AND EXPERIMENTATION

Engineering, as I have noted earlier, is rooted simultaneously in present-day designs—be they for a bridge, a road, or a spaceship—but done with future goals and projects in mind. Artifacts of the future, such as a space elevator or Drexlerian nano-assemblers, have existed in a similar state. Marked by the subjunctive—what might or could be, as opposed to what is or will be—they inhabit a liminal arena of possibility (Squier, 2004; Turner, 1977). Similarly, a diverse collection of people—from Nobel prize–winning scientists to basement mechanics—visited the same liminal space where they took advantage of the opportunity to "fantasize and play with new objects and possibilities—to dream, to model, and to tinker in new ways" (Pinch, 2007). Seen in this fashion, liminality can help us understand some key aspects of the space elevator, and indeed, it could well be used to describe other less visible aspects of nanotechnology's history.

An analogy drawn from the history of technology can also help us better apprehend the present/future existence of a nano-enabled space elevator. In the interwar period, space exploration was the primary "futurist technology" as researchers proposed theories of rocket propulsion and design on paper (Winter, 1983). Until the proper materials and institutional support became available, however, the rocket existed largely in a speculative state. This changed when amateur groups such as the *Verein für Raumschiffahrt* and the American Rocket Society helped propel the rocket from ridicule to reality. And for rocket gurus like Malina and von Braun, who were building actual hardware, science fiction remained an important inspiration.

Like the early rocket tinkerers, today's grassroots space elevator groups are typically situated outside of the mainstream of patronage for science and engineering, that is, they are not operating with multimillion-dollar NASA or NSF grants. Their activities reflect the myth of the lone inventor winning fortune and fame with technologies developed in garages or dorm rooms—think of Philo Farnsworth, Steve Jobs and Steve Wozniak, or Larry Page and Sergey Brin. No doubt, this linkage to a mythical part of America's technological past accounted for the favorable and widespread coverage given by the media to events like the Space Elevator Games. And, like the early rocket pioneers, space elevator advocates take inspiration from fiction and imagine that their work will pay off someday, perhaps with far-reaching societal consequences. The contributions of amateur scientists to the production of new knowledge have been well established by historians of science (McCray, 2008). Space elevator–themed contests are another example of how people outside the mainstream academic-corporate research community can help foster new innovations today just as they did for the personal computer in the 1970s.

Finally, the space elevator, situated at the confluence of nanotech and space aspirations, raises some questions about the imagined unity of converging technologies at some future point (Roco & Bainbridge, 2003). Although we can speak of it as a single entity, the reality is that nanotechnology—like space exploration before it—is a fragmented field. It encompasses a diverse range of activities, practices, goals, institutions, pedagogies, and instrumental techniques, making it more of a sociological than an epistemological phenomenon. The idea that nanotechnology will somehow fuse with biotechnology, information technology, and cognitive science and become a universal discipline does not reflect the messiness of actual technological innovation (see Bowker, 1993). What history shows is that technologies, even hypothetical ones, rarely converge into a single *über*-technology but are constantly converging, splitting, mutating, and reconverging as new ideas are proposed, tried, and adopted or discarded. Along the way, a host of diverse institutions and individuals, from a major NASA lab to groups of independent tinkerers to Nobel prize winners to futurists, used the space elevator concept in various ways to advance particular agendas.

Nonetheless, nanotechnology (specifically, the discovery of carbon nanotubes) was central to moving the space elevator to a stage somewhere between speculation, detailed design, and garage tinkering. Seen more broadly, the attention of science writers and other journalists on these new allotropes of carbon material stimulated huge amounts of attention as to what their possible applications might be. Smalley, a scientist predisposed to futurist imagining yet deeply rooted in mainstream lab-based research, helped make a space elevator, along with nanotechnology itself, seem credible. However, the discovery of carbon nanotubes would have had far less impact if nano-advocates like Drexler had not already promoted nanotechnology as the most promising technology of the twenty-first century. Together, the advocacy of nanotechnology, the invention of a new material with promising properties, detailed engineering studies by Edwards and others, and modest NASA support brought credibility to the idea. To top it all off, the interest and activity shown by amateurs and weekend engineers created a compelling frame for the media—that of David, with his "space elevator sling," taking on the corporate aerospace Goliaths—which, along with science-fiction narratives, brought the space elevator publicity and a global audience.

Technologies are the tools people use to make their future. Accompanying nuts-and-bolts hardware are less tangible tools such as ideas and plans for what the future should or could be like. Situated between fiction and reality for over a century, prize-winning science fiction, mainstream media attention, serious technology studies, commercial investment, and enthusiasm of amateur technologists have all nurtured the space elevator for decades. In order to understand the social life of emerging technologies, we must appreciate that histories of the technological future are of importance regardless of whether the technologies themselves ever become reality, let alone converge.

ACKNOWLEDGMENTS

This chapter is based upon work supported by the National Science Foundation under Grant No. SES 0531184. Any opinions, findings, and conclusions or recommendations expressed in this material are those of the author and do not necessarily reflect the views of the National Science Foundation. The original version was presented at the 2007 annual meeting of the Society for the History of Technology in Washington, DC. I wish to acknowledge contributions from people who assisted in the research for this essay including Mary Ingram-Waters and scientists and engineers who agreed to be interviewed.

NOTES

1. Vinge (1983) later acknowledged that the term originated, so far as he knew, in a tribute Stanislaw Ulam gave in 1958 for his colleague, mathematician John von Neumann.
2. At least this was the case a decade ago. More recently, discussions among science journalists and activists, who often act as bellwethers of change, have been discussing synthetic biology with increasing frequency; ex: http://www.synbioproject.org/about/
3. The book was originally published with a different subtitle—*Challenges and Choices of the Last Technological Revolution*—which hews closer to Drexler's original vision that his book be about technology in general rather than a treatise on nanotechnology per se.
4. In some ways, this is reminiscent of nanotech advocates citing Richard Feynman's 1959 "There's Plenty of Room at the Bottom" speech as the origin of their field.
5. Pearson (1997) concludes that although Tsiolkovsky anticipated the space elevator idea the actual credit for the invention should go to him and Yuri Artsutanov who independently published on the idea.
6. A case in point is the work done on a nuclear powered spacecraft. For example, Project Orion was a well-funded effort that attracted serious interest from physicists Ted Taylor and Freeman Dyson; it was disbanded in 1965 (Dyson, 1965).
7. "Unobtanium" is a colloquial term used in the aerospace industry to denote, as the name suggests, something which is too expensive or exists only in theory. More recently, corporate lust for the hypothetical material was a plot device used in the 2009 Oscar-winning movie *Avatar.*
8. Arthur C. Clarke and Charles Sheffield both published sci-fi novels featuring a space elevator in the late 1970s.
9. One reviewer of Edwards' proposal was Gentry Lee, a chief engineer at the Jet Propulsion Laboratory who co-authored several science fiction books with Arthur C. Clarke and was also a member of NIAC's Science, Exploration,

and Technology Council. According to Edwards, Gentry Lee "caught the proposal half way to the garbage can and got people to look at it more closely" (B. C. Edwards, personal communication, June 4, 2007). Edwards, according to NIAC's last report in 2007, also raised some $8.5 million in non-NASA funding.

10. According to the Executive Summary in Edwards' (2003) final report, the space elevator concept resulted in "hundreds of television, radio, and newspaper spots around the world," making it arguably the most visible project that NIAC funded (p. 2).
11. The idea of prizes for technological accomplishments has a long history going back to the celebrated eighteenth-century prize for determining longitude. In aerospace history, prizes have been particularly noteworthy. The Orteig Prize (for the first trans-Atlantic flight) and, more recently, the million-dollar Ansari X Prize (for the first nongovernment launch of a reusable, manned spacecraft, which the SpaceShipOne team won in 2004) are examples. Recently, there has been increased interest on the part of government agencies such as NASA, the Department of Energy, and DARPA to sponsor technology prizes (for one perspective see Kalil, 2006).

REFERENCES

Artsutanov, Y. (1960, July 31). V Kosmos na Electrovoze. *Komsomolskaya Pravda*.

Barry, D. (2003, November 9). Scientists launching spacey idea. *The Miami Herald*. Retrieved December 2011, from http://www.physics.ucsb.edu/~scipub/f2004/DaveBarry.pdf

Bostrom, N. (2005). A history of transhumanist thought. *Journal of Evolution and Technology, 14*(1), 1–25.

Bowker, G. (1993). How to be universal: Some cybernetic strategies, 1943–1970. *Social Studies of Science, 23*(1), 107–127.

Chang, K. (2003, September 23). Not science fiction: An elevator to space. *New York Times*, p. F1.

Chang, K. (2009, November 8). Winner in contest involving space elevator. *New York Times*. Retrieved January 2010, from http://www.nytimes.com/2009/11/08/science/space/08nasa.html

Clarke, A. C. (1979). *The fountains of paradise*. New York: Harcourt Brace Jovanovich.

Constant, E. (1980). *The origins of the turbojet revolution*. Baltimore, MD: The Johns Hopkins University Press.

Crouch, T. (1999). *Aiming for the stars: The dreamers and doers of the space age*. Washington, DC: Smithsonian Institution Press.

Drexler, K. E. (1986). *Engines of creation: The coming era of nanotechnology*. New York: Anchor Books.

Drexler, K. E. (1988). Exploring future technologies. In J. Brockman (Ed.), *The reality club* (pp. 129–150). New York: Lynx Books.

Dyson, F. J. (1965). Death of a project. *Science, 149*(3680), 141–144.

Edwards, B. (2003). The space elevator: NIAC phase II final report. Atlanta, GA: NASA Institute for Advanced Concepts.

Edwards, B., & Westling, E. (2003). *The space elevator: A revolutionary earth-to-space transportation system*. Houston, TX: BC Edwards.

Hassler, S. (2008, June). Un-assuming the singularity. *IEEE Spectrum*, p. 9. (entire issue is devoted to an examination of the singularity concept).

House of Representatives Committee on Science. (1999). *Nanotechnology: The state of nano-science and its prospects for the next decade*. Washington, DC: U.S. Government Printing Office.
Iijima, S. (1991). Helical microtubules of graphitic carbon. *Nature, 354*(6348), 56–58.
Interagency Working Group on Nanoscience, Engineering, and Technology. (1999). *Nanotechnology: Shaping the world atom by atom*. Washington, DC: National Science and Technology Council.
Isaacs, J. D., Vine, A. C., Bradner, H., & Bachus, G. E. (1966). Satellite elongation into a true "sky-hook." *Science, 151*(3711), 682–683.
Johnson, A. (2006). Institutions for simulations: The case of computational nanotechnology. *Science Studies, 19*(1), 35–51.
Johnson, A. (2010). *Hitting the brakes: Engineering design and the production of knowledge*. Durham, NC: Duke University Press.
Kalil, T. (2006). *Prizes for technological innovation*. Washington, DC: The Brookings Institution.
Kline, N. S., & Clynes, M. (1961). Drugs, space, and cybernetics: Evolution to
Cyborgs. In B. E. Flaherty (Ed.), *Psychophysiological aspects of space flight*, pp. 345–371. New York: Columbia University Press.
Kline, R. (2009). Where are the cyborgs in cybernetics? *Social Studies of Science, 39*(3), 331–362.
Kroto, H. W., Heath, J. R., O'Brien, S. C., Curl, R. F., & Smalley, R. E. (1985). C_{60}: Buckminsterfullerene. *Nature, 318*(6042), 62–63.
Kurzweil, R. (2005). *The singularity is near: When humans transcend biology*. New York: Viking.
Lvov, V. (1967). Sky-hook: Old idea. *Science, 158*(3803), 946–947.
Mazlish, B. (Ed.). (1965). *The railroad and the space program: An exploration in historical analogy*. Cambridge, MA: MIT Press.
McCray, W. P. (2005). Will small be beautiful? Making policies for our nanotech future. *History and Technology, 21*(2), 177–203.
McCray, W. P. (2008). *Keep watching the skies! The story of operation moonwatch and the dawn of the space age*. Princeton, NJ: Princeton University Press.
McCurdy, H. E. (1997). *Space and the American imagination*. Washington, DC: Smithsonian Institution Press.
McMaster, J. (Producer). (2007, January 9). *Nova: Space elevator* [Television broadcast]. Boston, MA: WGBH for PBS.
Misra, M. (1990). Towards unobtainium: New composite materials for space applications. *Aerospace Composites and Materials, 2*(6), 29–32.
Mody, C. (2010). *Institutions as stepping stones: Rick Smalley and the commercialization of nanotubes*. Philadelphia, PA: Chemical Heritage Foundation.
Moravec, H. (1977). A non-synchronous orbital skyhook. *The Journal of the Astronautical Sciences, 25*(4), 307–322.
National Academy of Engineering. (1999). Concerning federally sponsored inducement prizes in engineering and science. Washington, DC: National Academy of Engineering.
NASA Institute for Advanced Concepts (NIAC). (2007, June 8). Long-term success of NIAC-funded concepts. Retrieved December 2011, from http://www.niac.usra.edu/files/misc/NIAC_ROI.pdf
Nordmann, A. (2004). Nanotechnology's worldview: New space for old cosmologies. *IEEE Technology and Society Magazine, Winter*, 48–54.
Pearson, J. (1975). The orbital tower: A spacecraft launcher using the earth's rotational energy. *Acta Astronautica, 2*, 785–799.
Pearson, J. (1997). *Konstantin Tsiolkovski and the origin of the space elevator*. Presented at the 48th International Astronautical Congress. October 6–10, Turin, Italy.

Pinch, T. (2007, October 18). *Between technology and music: Distributed creativity and liminal spaces in the making and selling of synthesizers.* Presented at the SHOT Workshop on the Animating Passions for the History of Technology, Washington, DC.

Robinson, K. S. (1993). *Red Mars.* New York: Bantam Books.

Roco, M. C., & Bainbridge, W. S. (Eds.). (2003). *Converging technologies for improving human performance: Nanotechnology, biotechnology, information technology and cognitive science.* London: Kluwer Academic Publishers.

Rynin, N. A. (1971). *K. E. Tsiolkovskii: Life, writings, and rockets: Interplanetary flight and communication,* Volume 3, No. 7; original work published in Leningrad, 1931; translated from Russian in 1971 by Israel Program for Scientific Translation and published for NASA and the National Science Foundation by the Israel program for Scientific translation.

Segal, H. (1985). *Technological utopianism in American culture.* Chicago: University of Chicago Press.

Smalley, R. (1993a). 21 January 1993 letter to M. Carroll and James Kinsey; Smalley papers, Box 3, Folder 3, Rice University Archives.

Smalley, R. (1993b). 5 October 1993 memo to J. Bourne; Richard Smalley papers, Chemical Heritage Foundation.

Smalley, R. (1996). 24 August 1996 email to Gustavo Scuseria; Box 59, Folder 7, Richard Smalley papers, Chemical Heritage Foundation.

Squier, S. (2004). *Liminal lives: Imagining the human at the frontiers of biomedicine.* Durham, NC: Duke University Press.

Thompson, R. (2009). *Structures of change in the mechanical age: Technological innovation in the United States, 1790–1865.* Baltimore, MD: Johns Hopkins University Press.

Turner, V. (1977). Frame, flow, and reflection: Ritual and drama as public liminality. In M. Benamou & C. Caramello (Eds.), *Performance in postmodern culture*, pp. 33–55. Madison, WI: Coda Press.

Vincenti, W. (1990). *What engineers know and how they know it.* Baltimore, MD: Johns Hopkins Press.

Vinge, V. (1983, January). First word. *Omni,* 10.

Winter, F. (1983). *Prelude to the space age: The rocket societies.* Washington, DC: Smithsonian Institution Press.

Yakobson, B., & Smalley, R. (1997, July–August). Fullerene nanotubes: $carbon_{1,000,000}$ and beyond. *American Scientist,* 324–337.

4 Conferences and the Emergence of Nanoscience

Cyrus C. M. Mody

There are certain activities that absorb scientists' time, yet which historians, sociologists, and (especially) philosophers of science largely neglect: writing grant applications, managing subordinates, convening committees, traveling, and so on. Analysts of science, at least since Kuhn, have instead focused almost exclusively on the explicitly knowledge-oriented characteristics that ostensibly set scientific practice apart from other occupations: scientists' "inscriptions" and journal articles, their public and private debates over knowledge, their metaphysical predispositions, their theoretical and experimental techniques for apprehending the world. In understanding scientists primarily as knowledge producers, analysts generally ignore mundane activities scientists have in common with other professionals.

One under-analyzed activity that occupies a great deal of many scientists' time is conference-going. There are, to be sure, a few mentions of conferences in the history and sociology of science literature. For instance, Ochs and Jacoby (1997) have shown how the conference calendar forces scientists to concretize their knowledge in a form that will persuade an audience. A few particular conferences have been acknowledged as important to the development of certain fields, such as the Macy Foundation workshops in cybernetics (Heims, 1993), the Asilomar meeting in molecular biology (Hindmarsh & Gottweis, 2005), or the competing American Chemical Society and American Physical Society conferences in the cold fusion controversy (Pinch, 1995). Some influential conference series have even generated their own histories, such as the Solvay (Mehra, 1975) and Pugwash (Rotblat, 1972) meetings. There have also been limited quantitative studies of the role of conferences in building social networks in science (Martins, Gonçalves, Laender, & Ziviani, 2009; Melin, 2000).

However, we know little about conferences *as institutions*: why and how scientists start, attend, or organize conferences; how conferences have evolved; what role they play in scientific identity or community formation; what happens in the vendors' exhibits that accompany (and bankroll) many scientific conferences; or how conferences establish conduits for moving people, techniques, and ideas. This lack of attention is odd given the past three decades' interest in the "social construction" of scientific knowledge.

Conferences are where scientists are at their most social, and where competing scientific agendas meet (almost literally) face-to-face.

Conferences are an especially promising analytical site for social studies of nanoscience and other emerging fields. An "emerging" area has few or weak institutions, a sparsely connected network of participants, and a shaky sense of common purpose or identity. Conferences are incredibly effective mechanisms for creating connections among participants and aligning them toward a shared identity or project. This is especially so for emerging areas not yet served by established institutions such as journals or professional societies.

All these conditions apply to nanotechnology. Nanotechnology is notoriously difficult to define; it took 10–15 years before strong institutions championed it, and its practitioners are still fragmented and rarely self-identify as "nanotechnologists." Conferences were a crucial tool of nano's supporters in weaving disparate fields together. Conversely, attaching the "nano" label to their communities' conferences allowed various research fields to overcome demographic decline and intellectual stagnation in the early 1990s.

Today, dozens of conference series have "nano" in their names. Probably half a dozen have ambitions to be *the* disciplinary conference for the nano enterprise. Yet because each of these conferences has its roots in different fields, none can claim to represent the whole field in the way that, say, the American Physical Society does for American physicists. Nano conferences are both centripetal mechanisms, bringing diverse practitioners under the nano umbrella, and centrifugal mechanisms, balkanizing practitioners into social worlds that are only partially permeable to each other.

This chapter examines the centripetal and centrifugal nature of conferences in two nano subfields: microfabrication and probe microscopy. These fields began as distinct specialties with little overlap. Conferences were critical to bringing both these fields into being in the late 1970s and early 1980s. In the early 1990s, adoption of the "nano" label by those conferences was instrumental in blurring the boundaries between these fields. By examining the evolution of the conference series associated with these fields, we gain a better sense of how nanotechnology became a mutual organizing principle for—and invitation for contact between—research communities that initially had little to do with each other. That, in turn, offers insight on the strengths and weaknesses of the "nanotechnology" label as an instrument of science policy.

More tentatively, this chapter will also sketch the outlines of a general analysis of conferences as knowledge-making institutions. Conference-going should be seen not as something scientists just happen to do, but as an indispensable tool for the creation, validation, and spread of the knowledge and practices that are the center of the scientific life. Constructivist historians and sociologists of science (and their intellectual neighbors) have long been interested in the epistemic role of more-or-less discrete social formations such as "core sets" (Collins, 1985) or closely-grouped nodes in "actor-networks" (Latour, 1987), yet they lack a language for describing social formations such as conferences that can have a formalized, enduring presence and visibility. Organizational sociologists, in comparison, possess a sophisticated language for describing how institutions such as conferences can acquire formal, enduring (but not static) characteristics and provide conduits for the spread of professional norms and practices (DiMaggio, 1991; DiMaggio & Powell, 1983). It should be obvious that scientists continually interact with formal, durable (if evolving) social formations; the language of organizational sociology should, therefore, be applicable to the sociology of scientific knowledge. Conferences are just the kind of institution that should be amenable to such an analysis. This chapter will make an initial foray in that direction by offering a rudimentary taxonomy of different kinds of conferences, and by outlining some of the different ways conferences emerge, evolve, and attach to research communities.

MICROFABRICATION

Microfabrication is the art and science of making tiny, usually functional, structures. That is a loose definition; more restrictive definitions are possible, but microfabrication is today such a big-tent field that the loose definition is the only one that covers most of the people who publish in microfabrication journals or attend microfabrication conferences. When the field emerged in the late 1960s, however, its jurisdiction (Abbott, 1988) was narrower; practitioners focused on using beams of radiation or particles to carve micron- or submicron-scale, crystalline, inorganic structures like those used in microelectronic or microelectromechanical devices. Thus, even today, the largest microfabrication conference series is known colloquially as the "Three Beams" meeting, after the ion, electron, and photon beams with which the first generation of microfabrication specialists carved small features.

Although individual conference series devoted to one or two of the three beams originated in the late 1950s, a joint electron, ion, and photon beam meeting only emerged in 1975 (Schattenburg, 2007). It was about this time, too, that organizers of the Three Beams meeting began peer-reviewing submitted papers and then publishing them in the well-respected *Journal of Vacuum Science and Technology*, the journal of the American Vacuum Society. For the Three Beams' predecessor conferences, papers had been unreviewed and self-published by the organizers in a limited-circulation proceedings. Thus, the mid-1970s marked a major turning point in the professionalization of microfabrication.

The mid-1970s also saw the Japanese government's announcement of a major initiative in very large-scale integrated circuits (VLSI). American firms and government agencies, apprehensive of Japanese microelectronics firms' achieving parity with U.S. competitors, were shaken by the VLSI initiative. "Institutional entrepreneurs" (Choi & Mody, 2009; Misa, 1985) within those organizations leveraged that panic to funnel resources into microfabrication research. Those new resources, in turn, facilitated the creation of formal, durable institutions for a microfabrication research community.

Jay Harris, a program officer for electrical engineering in the National Science Foundation (NSF) Engineering Division, and his boss, Tom Meloy, were two such entrepreneurs. Upon learning that the Engineering Division was looking for a marquee project to put engineering on a par with other divisions in the NSF, Harris—with Meloy's support—proposed funding a national microfabrication facility (Harris, 2004). However, the NSF's governing body, the National Science Board (NSB), worried that such a facility might compete with industry and that oversight on a large block grant to establish and run the facility might be untenable.

Before approving Harris' proposal, therefore, the NSB demanded evidence of support from academic and industrial researchers. Accordingly, Harris organized three workshops in May 1976 to gauge opinion. Because Harris influenced the attendee list, it is unsurprising that participants were generally positive. Still, even participants who were initially hostile left with a change of heart (R. F. W. Pease, personal communication, November 16, 2005). Any general sociological or historical study of scientific conferences needs to include *ad hoc* workshops like Harris' that reify a research community's opinion or call that community into being. *Ad hoc* workshops can evolve into lasting institutions, or set the agenda for permanent conference series.

In the case of microfabrication, Harris' workshops coincided with the start of a biennial Gordon Research Conference (GRC) series on the "Chemistry and Physics of Microstructure Fabrication." The instigator of this series was Robert Keyes, an expert on the effects of miniaturization and integration on electron device characteristics at IBM. Keyes was anxious to put American microfabrication research on a scientific basis, rather than let it continue as a cut-and-try endeavor (Keyes, 1975). Yet, perhaps ironically, he felt that doing so required a cozy, *gemeinschaftlich* setting that the increasingly formal, impersonal, *gesellschaftlich* Three Beams meeting had abandoned. Again, any general analysis of scientific conferences should explore the complementarity of such institutional poles: formal and informal, personal and impersonal, large and small.

The GRC series was a perfect forum for such a personalized meeting. The "Gordons" began in 1931 as small conferences in cutting-edge research areas (Daemmrich, Gray, & Shaper, 2006), and by 1975 had grown to roughly 90 conferences a year. From the beginning, all the Gordon conferences have shared quixotic features inspired by scientific utopian ideals. Conferences were long (usually four full days, with participants pressured to stay the whole time), remote (most met at small boarding schools in rural New England), invitation only, small (never more than 150 people, often only 30 or 40), off the record (no proceedings or talks were to be cited in the published literature), and cutting edge (talks were to cover material never presented before). Oddest of all was the schedule. Two or three plenary talks filled each morning and evening. In between, participants did whatever they wanted—there was no scheduled activity. For Keyes, this made the GRC format ideal for an emerging microfabrication community which needed some off-the-record way to exchange ideas and unpublished tricks for making tiny devices. That kind of communication required trust that could be built more easily during a sail on Lake Winnipesaukee or a hike through the New Hampshire woods than through formal, scheduled activities.

Harris' workshops and NSF's coming call for microfabrication facility proposals loomed over the first Microfabrication Gordon Conference in 1976. All the top contenders in the NSF competition—Berkeley, MIT/Lincoln Labs, and Cornell—were represented on Keyes' speaker list. Ed Wolf from Hughes Research was Keyes' vice-chair and adviser on Berkeley's proposal; when Cornell won the competition, he became director of the facility. Other speakers included: Tom Everhart, who led Berkeley's submission and then became Cornell's Dean of Engineering specifically to oversee the facility; Hank Smith of Lincoln Labs, who led MIT's proposal and then became director of MIT's privately-funded facility built in response to NSF's decision to fund Cornell; and Noel MacDonald, then of Physical Electronics, who was one of the first people Wolf and Everhart lured to work at the Cornell facility. Of the 20 speakers at the 1976 Gordon Microfabrication Conference, those 4 soon affiliated with Cornell or MIT, and a further 8 were at IBM or Bell Labs. Well into the 1990s, almost all the chairs of the Gordon Microfabrication Conference were currently or recently affiliated with the Cornell or MIT facilities, IBM, or Bell Labs. Several chairs had spent time at more than one of those organizations.

That alone reveals something about the social and intellectual world of microfabrication. Through the 1990s, East Coast institutions dominated the Gordon Microfabrication meeting. West Coast firms were largely absent; for instance, I can find only two attendees each from Intel and Fairchild Semiconductor and almost none from other Silicon Valley firms between 1976 and 1998, compared to about 60 each from AT&T (or its daughter companies) and IBM. Leading West Coast universities were also less well represented: about a half dozen attendees each from Berkeley, Caltech, and the University of California Santa Barbara (UCSB), and a dozen from Stanford, compared to about 20 each from MIT and Harvard and 40 from Cornell. Moreover, UCSB and Stanford were drawn into the GRC series only *after* a prominent microfabrication specialist and eventual chair of the GRC meeting moved to each school from Bell Labs—Fabian Pease to Stanford in 1978 and Evelyn Hu to UCSB in 1984.

From this we can surmise a geographical split in the microfabrication community, particularly between Silicon Valley and the Northeast. True, GRC meetings in New Hampshire were more accessible to Northeasterners; yet distance did not discourage attendees from Texas or even the UK and

Japan. A more likely explanation for the split is that the dominant Silicon Valley firms were, by the late 1970s, operating with a much shorter time horizon from research to commercialization than IBM and AT&T. Intel and its peers were thus less interested in the esoteric lithographies and materials discussed at the GRC. Stanford's microfabrication facility—the leading academic microfabrication center on the West Coast—had a more mature industrial affiliates program (Lécuyer, 2005) than Cornell's or MIT's. Firms that wanted extended face-to-face interaction with Stanford researchers could become industrial affiliates, whereas the GRC may have been a better venue for firms to pick the brains of MIT and Cornell people.

Critically, this geographical split was not static. Again, the GRC meeting's content and personnel help us trace the evolution of ideas and flows of personnel in the microfabrication community. Over the course of the 1980s, East Coast research labs gradually declined, then precipitously contracted in the 1990s. Many corporate microfabrication specialists left for university positions, where they drew students and other faculty members to the conference with them. Thus, the share of attendees employed by industry went from 77 percent in 1976 to only 16 percent in 1998.

Waning corporate participation triggered a crisis in the microfabrication series. Silicon Valley firms' adherence to optical lithography (dismissing the other "beams"), and their adoption of larger, more expensive wafers, meant that the relevance of academic practitioners' expertise to the microelectronics industry was attenuated. Academic microfabrication specialists needed new audiences outside the microelectronics industry. Given the appeal of the biotech industry in the 1980s and skyrocketing National Institutes of Health (NIH) budgets in the 1990s, biology and medicine proved especially popular as new outlets for microfabrication techniques. In moving toward such applications, however, microfabrication specialists could not be sure that the traditional "three beams" were the most appropriate tools. Specialists increasingly became interested in novel fabrication techniques as complements or alternatives to ion, photon, and electron beams.

The GRC home office sensed as early as 1988 that the conference leadership was not responding to these changes, warning that "there might be an ingrowing clique forming which might be deleterious to branching out with significant representation in not-so-well-known areas" (Conference on Microstructure Fabrication, 1988). Because such a warning could lead to cancellation of the series, the organizers of the 1992 conference revamped it to include topics such as lithography using a scanning tunneling microscope. The next meeting, in 1994, saw an even more dramatic expansion to include topics such as "Tracking down Biological Motors Using Optical Tweezers," "Microfabricated Arrays: DNA Electrophoresis and Cell Mobility," and "Biocatalytic Synthesis of Polymers of Precisely Defined Structure."

But was this really "microfabrication" anymore? Certainly, broadening the conference's focus attracted attendees with less loyalty to the traditional tools and aims of microfabrication. So at the 1994 meeting, attendees voted to change the series name to "Chemistry and Physics of Nanostructure Fabrication." Why nano? The simple reason is that microfabrication techniques had improved, and describing feature sizes in billionths ("nano") rather than millionths ("micro") of a meter signaled that progress. Yet there were likely additional considerations. Even in 1994, "nanotechnology" was a widely circulating label for a broad spectrum of research. Adopting the nano label made it easier for leaders of the GRC series to justify expansion beyond the three beams—or, at least, "nanofabrication" more accurately reflected the conference's new, ecumenical focus than "microfabrication" did.

At the time, other conferences, journals, and centers were adopting the nano label. Following suit made it easier for the GRC series to attract affiliates of those other institutions. For instance, the Cornell, MIT, and Stanford fabrication facilities that supplied so much of the GRC meeting's attendance all incorporated "nano" in their names by the early 1990s. Likewise, the Microcircuit Engineering Conference changed its name to Micro and Nano Engineering International Conference in 1994, and the Three Beams meeting became the International Conference on Electron, Ion, and Photon Beam Technology and Nanofabrication in 1995. To see when "nano" came into being and how it spread, conferences are an excellent tracer.

PROBE MICROSCOPY

Microfabrication is old-line nano, based on old techniques that became "nano" over the course of the 1980s. Thus, it often receives less attention than it merits in histories of nanotechnology—for a supposedly novel field to include such an ancient practice could be embarrassing. This was not the case with probe microscopy, which was invented in the early 1980s and found itself at the nanoscale from the beginning (Mody, 2011). Explanations of what nano is and where it came from almost always gesture to probe microscopy. Yet it was not always self-evident to probe microscopists that what they were doing was nanotechnology. Indeed, some were resistant to that label well into the 1990s. Thus, the "standard story" (Baird & Shew, 2004) in which probe microscopy is foundational to nanotechnology ignores the slow, difficult work needed to make probe microscopy self-evidently "nanotechnological." Conferences, it turns out, were critical in that process.

Indeed, the probe microscopy community began with a conference. Although invented in 1981 at the IBM Zurich research lab, the technique spread slowly until 1985. Most early adopters, especially in the US, faced long delays getting their microscopes to work until one of them, Stanford's Calvin Quate, convened an *ad hoc* workshop in Cancun in late 1984. There, attendees quizzed one of the technique's inventors, Heini Rohrer, and each other for practical advice (e.g., on where to buy various parts).

Within weeks of the Cancun meeting, and possibly because of it, several attendees successfully replicated the technique, which they demonstrated to a standing-room-only audience at the March 1985 American Physical Society meeting. With interest in probe microscopy growing rapidly, Rohrer organized another one-off workshop later that year. This meeting, sponsored by IBM Europe and held in Oberlech, Austria, featured most of the Cancun group, plus other early adopters from Europe and North America. Where the Cancun meeting was fraught with anxiety, the Oberlech meeting was boisterously good-natured. Probe microscopists look back at this conference with exceptionally happy thoughts, especially about friendships they struck with other early adopters (S. Chiang, personal communication, March 8, 2001; Gimzewski, personal communication, October 22, 2001).

One attendee, Nico García, a Spanish academic with ties to IBM Zurich, proposed forming a permanent International Conference on Scanning Tunneling Microscopy. In 1986 he hosted the first "STM Conference" (as it became known) in Santiago de Campostela. From this meeting emerged an organizing committee that annually steered the conference to regions where it could catalyze local interest in the technique and thereby expand the probe microscopy community: California in 1987, the UK in 1988, Japan in 1989, and the U.S. East Coast in 1990.

This committee also succeeded in getting the conference proceedings published in the *Journal of Vacuum Science and Technology* (JVST), the same journal that published proceedings of the Three Beams meeting, and in affiliating the STM Conference with JVST's parent organization, the American Vacuum Society (AVS). This was largely the work of James Murday, a grant officer at the Office of Naval Research (ONR) and surface chemist at the Naval Research Lab (NRL). Murday was influential in the AVS (he was its president in 1992); he was also well-known in probe microscopy, both because he ran an STM group at NRL and because he oversaw ONR grants to probe microscopy researchers.

As organizer of the 1990 STM Conference, Murday tackled an emerging demographic and organizational problem for probe microscopists. When the technique was invented, it was limited to scanning tunneling microscopy (STM)—hence the "STM Conference." However, in 1985 one of the STM's inventors, Gerd Binnig, along with Binnig's longtime technician, Christoph Gerber, and Stanford's Cal Quate, co-invented a variant, the atomic force microscope (AFM). More probe techniques soon followed. By 2001, there were close to 50 named probe microscopies, and the STM Conference had morphed, at least officially, into the International Conference on Scanning Tunneling Microscopy/Spectroscopy and Related Techniques.

Even in 1990, the "STM Conference" was not just an STM conference. Some STMers, especially those who studied semiconductor and metal surfaces under ultrahigh vacuum (UHV) conditions, considered research done with the new variants—AFM and STM in air, in liquid, or to study biological materials—to be of inferior quality. Problematically, UHV STMers were closer to the AVS's core membership than practitioners of the new variants. Murday, however, wanted the newcomers to affiliate with AVS, seeing them as potentially reinvigorating the Society.

Thus, he refashioned the 1990 meeting as the Fifth International Conference on Scanning Tunneling Microscopy/Spectroscopy/First International Conference on Nanometer Scale Science and Technology (or NANO Conference). He hoped that the NANO meeting would soon replace the STM Conference. As he puts it, however, probe microscopists were "not quite ready" for that (J. Murday, personal communication, May 29, 2007). So as president of AVS in 1992, Murday arranged a compromise: NANO would run in even years, the STM Conference in odd years. The two would have overlapping attendance and content; NANO would be larger and broader but with probe microscopy at its core. This arrangement worked satisfactorily until 2006, when the two finally merged as the International Conference on Nanoscience + Technology.

Some UHV STMers, however, saw a different solution to the STM Conference's demographic and intellectual frictions. Uncomfortable sharing a conference with other kinds of probe microscopists, they tried to form a breakaway Gordon Conference in 1993. As the GRC home office noted, "This new conference is essentially a physics-physical chemistry splinter group of STM practitioners. They found the international STM conference too big and impersonal and wanted to distance themselves from the biological STM community whose work they do not hold in high regard. This resulted in a very small conference" (Conference on Scanning Tunneling Microscopy, 1993). Even some of these STM secessionists agreed with the GRC that the focus was too narrow. One attendee commented in his/her GRC survey that

> [t]he topics covered represented only a portion of the frontiers in S.T.M.. [*sic*] Noticeably missing were presentations from biological S.T.M. as well as workers in the area of thin films. This conference was completely devoted to S.T.M. in U.H.V.. [*sic*] Future conferences should include, not exclude, contributions from these areas.

In the end, the GRC organization concluded that the conference series should be canceled after its first meeting.

NANO PROLIFERATION

Even if one accepted the premise that probe microscopy's expansion diluted its quality, by the early 1990s the genie was out of the bottle. Still undetermined, however, was whether "nano" was an appropriate label uniting the diverse variants and applications of probe microscopy and connecting them with fields such as microfabrication. Murday and other nano supporters spent the 1990s attaching that label to an increasing number of institutions, each reciprocally legitimating the others by giving the sense that (as Murday puts it) the "freight train" of nanoscale science and technology was rushing forward.

Conferences were not the only institutions that "nanoized" in the 1990s. Indeed, conference series adopted the nano label in feedback with other kinds of institutions. We have seen that the Three Beams and GRC Microfabrication meetings adopted "nano" in tandem with similar name changes at the academic nanofabrication facilities that contributed many of their attendees. Similarly, Murday assembled a series of nano conferences in the late 1980s and early 1990s that prepared the way for further nano conferences and institutions. For instance, he organized AVS topical conferences on Nanometer Scale Properties of Surfaces and Interfaces in 1988 and 1989, partly in order to create support for establishing the NANO meeting in 1990. He then used the NANO series to justify creating a Nanometer Scale Science and Technology Division within AVS in 1992. That year he also organized a conference on Atomic and Nanoscale Modification of Materials for the Engineering Foundation and, in 1995, became the first chair of a Nanometer Structures Division in AVS's international umbrella organization, the International Union for Vacuum Science, Technique, and Applications.

Through the proliferation of nano conferences and other institutions in the early 1990s, probe microscopists and other scientists were repeatedly exposed to the nano label and eventually internalized it enough to become advocates themselves. An important example of that internalization can be glimpsed in the 1992 travel schedule of Rick Smalley, the Nobel laureate and co-discoverer of buckminsterfullerenes. As other contributions to this volume (Eisler, Chapter 2; McCray, Chapter 3) show, Smalley was one of the most influential advocates for the formation of a National Nanotechnology Initiative. He was also one of the earliest and most energetic creators of "nano" institutions (Mody, 2010). Yet Smalley's support for "nano" was not inevitable—it built incrementally, in part through repeated exposure to the term at conferences.

In fact, it is unclear whether Smalley had even heard of nanotechnology before 1992; he certainly was not strongly attached to the term. So one of his first exposures came in January that year, when he went to a conference in Japan on "Nanotechnology: Science at the Atomic Scale," organized by editors at *Nature*. In May, he attended a workshop on nanotechnology in Santa Fe, New Mexico, probably organized by defense research policymakers. In August, he went to Murday's Engineering Foundation conference. By September, he was referring to his own work as "fullerene nanotechnology" at another conference in Japan. Finally, in October, the Rice chemistry department held a retreat to discuss its plans for the next decade. There, Smalley argued for reorienting the department toward nanotechnology and biochemistry. A few months later he testified before Congress on the need for funding and coordination to enable development of "practical, useful nanotechnology" (Byerly, 1993).

Obviously, the roots of Smalley's advocacy for nanotechnology cannot be traced just to the proliferation of nano conferences in the early 1990s. Moreover, whatever influence those conferences had on Smalley could not have been due just to the "nano" prefix in their titles; there had to be content in those conferences that made the argument for nanotechnology persuasive. Yet we can pick out some reasons why conferences were an appealing tool for making that argument and drawing influential scientists such as Smalley, as well as many probe microscopists and microfabrication specialists, to the cause of nanotechnology.

Most importantly, conferences offer massive co-presence (Beaulieu, 2010; Goffman, 1967). They bring large numbers of people together who then share knowledge, develop connections (leading to later jobs, collaborations, or funding), and build trust that inclines participants to notice commonalities not apparent from textual sources alone. Just by the way its program, committees, and invitation lists were constructed, a nano conference in the 1990s could make an argument for the legitimacy of "nano" as an organizing principle for research. For instance, by seeing probe microscopists, fullerene chemists, and dendrimer researchers together on a panel, conference attendees would be guided to appreciate how "nano" connects those fields.

Other institutions (e.g., research centers) also afford co-presence in ways that argue for nano as a legitimate organizing principle. Several features set conferences apart from other instruments of co-presence, however. First, conferences have less overhead—financial, temporal, intellectual—than other institutions. An *ad hoc* conference can focus attention on a topic with little investment of time or resources. Even if that *ad hoc* meeting evolves into something more permanent, a conference series usually does not involve the expense of a brick-and-mortar research center. Conferences' low overhead means they can be more experimental. *Ad hoc* conferences can try-out new ways of organizing a field to see whether participants will commit to it. That commitment never materialized for the GRC STM conference and took 15 years to solidify for the NANO meeting.

Conversely, conferences foster attachment to a set of ideals because they are *not* completely without overhead—they require individuals and organizations to commit time and resources to organize and attend them. That commitment can lead participants to internalize the conference's organizing principles. Because a conference in a series is usually organized by different people each time it meets, the circle of people committed to that conference's principles continually grows. Conference series also *evolve* as different people take their turn assembling its program and invitee list. Conferences with closed sets of organizers or attendees die out over time—hence the GRC's frustration with its microfabrication conference in the late 1980s. Conferences that continually seek commitment from new groups can cultivate a larger constituency and respond to a changing environment—although they risk losing a tight topical focus (a common complaint about many nano meetings today).

Finally, conferences are effective tools for guiding science because they are attached to, or serve as bridges toward, other types of institutions. One set of institutions, for instance, might supply a conference's attendees, whereas another might offer logistical support or the legitimacy of peer-reviewed publication for its proceedings. Whether to commit to a novel organizing principle like nano is a question for institutions, such as professional societies, as much as for individuals. Conferences, again, can offer a way to experiment with that commitment. Thus, conferences can foster change within institutions. In probe microscopy, for instance, AVS's commitment to the NANO conference gradually broadened the focus and membership of the society itself. Similarly, the expanding focus of the GRC micro/nanofabrication series evolved in feedback with the broadening focus of micro/nanofabrication facilities at Cornell, MIT, Stanford, and elsewhere. If we want to understand how nanotechnology has effected change within scientific communities and institutions, we need to attend to the ways conferences have facilitated those changes. More generally, if we want to understand how communities shape scientific knowledge and practice, we need to pay attention to institutions such as conferences that allow individuals to come together and ratify their membership in such communities.

ACKNOWLEDGMENTS

This material is based upon work supported by the National Science Foundation under Cooperative Agreements Nos. SES 0531184 and SES 0938099 to the Center for Nanotechnology in Society at University of California at Santa Barbara. Any opinions, findings, and conclusions or recommendations expressed in this material are those of the authors and do not necessarily reflect the views of the National Science Foundation.

REFERENCES

Abbott, A. D. (1988). *The system of professions: An essay on the division of expert labor.* Chicago: University of Chicago Press.

Baird, D., & Shew, A. (2004). Probing the history of scanning tunneling microscopy. In D. Baird, A. Nordmann, & J. Schummer (Eds.), *Discovering the nanoscale*, pp. 145–156. Amsterdam: IOS Press.

Beaulieu, A. (2010). From co-location to co-presence: Shifts in the use of ethnography for the study of knowledge. *Social Studies of Science, 40*(3), 453–470.

Byerly, R., Jr. (1993). Letter from Radford Byerly, Jr. (chief of staff, US House of Representatives, Committee on Science, Space, and Technology) to Rick Smalley, March 1, 1993, Richard Smalley Papers, 1990–1998, MS #490, Woodson Research Center, Fondren Library, Rice University, Box 2, Folder 49.

Choi, H., & Mody, C. C. M. (2009). The long history of molecular electronics: Microelectronics origins of nanotechnology. *Social Studies of Science, 39*(1), 11–50.

Collins, H. M. (1985). *Changing order: Replication and induction in scientific practice*. London: Sage.

Conference on Microstructure Fabrication. (1988). Written comments made by conferees 1988. Gordon Research Conference papers, Chemical Heritage Foundation, Series VI (Evaluations), Box 129 (1983–1989).

Conference on Scanning Tunneling Microscopy. (1993). 1993 conference evaluation. Gordon Research Conference papers, Chemical Heritage Foundation, Series VI (Evaluations), Box 131 (1993–4).

Daemmrich, A. A., Gray, N. R., & Shaper, L. (Eds.). (2006). *Reflections from the frontiers, explorations for the future: Gordon Research Conferences, 1931–2006*. Philadelphia: Chemical Heritage Press.

DiMaggio, P. J. (1991). Constructing an organizational field as a professional project: U.S. Art Museums, 1920–1940. In W. W. Powell & P. J. DiMaggio (Eds.), *The new institutionalism in organizational analysis*, pp. 267–292. Chicago: University of Chicago Press.

DiMaggio, P. J., & Powell, W. W. (1983). The iron cage revisited: Institutional isomorphism and collective rationality in organizational fields. *American Sociological Review, 48*(2), 147–160.

Goffman, E. (1967). *Interaction ritual: Essays on face-to-face behavior*. Garden City, NY: Anchor Books.

Harris, J. H. (2004). *It's a small world*. Unpublished paper for the 25th anniversary of the Cornell Nanofabrication Facility.

Heims, S. J. (1993). *Constructing a social science for postwar America: The cybernetics group, 1946–1953*. Cambridge, MA: MIT Press.

Hindmarsh, R., & Gottweis, H. (2005). Recombinant regulation: The Asilomar legacy 30 years on. *Science as Culture, 14*(4), 299–307.

Keyes, R. W. (1975). Gordon Research Conferences conference proposal. Gordon Research Conferences papers, Series III, Box 51, Folder "Microstructure Fabrication," Chemical Heritage Foundation.

Latour, B. (1987). *Science in action: How to follow scientists and engineers through society*. Cambridge, MA: Harvard University Press.

Lécuyer, C. (2005). What do universities really owe industry? The case of solid state electronics at Stanford. *Minerva, 43*(1), 51–71.

Martins, W. S., Gonçalves, M. A., Laender, A. H. F., & Ziviani, N. (2009). Assessing the quality of scientific conferences based on bibliographic citations. *Scientometrics, 83*(1), 133–155.

Mehra, J. (1975). *The Solvay conferences on physics: Aspects of the development of physics since 1911*. Dordrecht, Holland: D. Reidel.

Melin, G. (2000). Pragmatism and self-organization: Research collaboration on the individual level. *Research Policy, 29*(1), 31–40.

Misa, T. J. (1985). Military needs, commercial realities, and the development of the transistor, 1948–1958. In M. R. Smith (Ed.), *Military enterprise and technological change: Perspectives on the American experience*, pp. 253–287. Cambridge, MA: MIT Press.

Mody, C. C. M. (2010). *Institutions as stepping-stones: Rick Smalley and the commercialization of nanotubes*. Philadelphia, PA: Chemical Heritag e Foundation.

Mody, C. C. M. (2011). *Instrumental community: Probe microscopy and the path to nanotechnology*. Cambridge, MA: MIT Press.

Ochs, E., & Jacoby, S. (1997). Down to the wire: The cultural clock of physicists and the discourse of consensus. *Language in Society, 26*(4), 479–505.

Pinch, T. J. (1995). Rhetoric and the cold fusion controversy: From the chemists' Woodstock to the physicists' Altamont. In H. Krips, J. E. McGuire, & T. Melia

(Eds.), *Science, reason, and rhetoric*, pp. 153–176. Pittsburgh, PA: University of Pittsburgh Press.

Rotblat, J. (1972). *Scientists in the quest for peace: A history of the Pugwash Conferences*, Cambridge, MA: MIT Press.

Schattenburg, M. L. (2007). *History of the "Three Beams" conference, the birth of the information age and the era of the lithography wars*. International Conference on Electron, Ion, and Photon Beam Technology and Nanofabrication. Retrieved March 8, 2012, from http://eipbn.org/wp-content/uploads/2010/01/EIPBN_history.pdf

Part II

Controlling the Field

The Role of Public Policies, Market Systems, Scientific Labor, and Globalization in Nanotechnology

5 Is Nanoscale Collaboration Meeting Nanotechnology's Social Challenge?

A Call for Nano-Normalcy

Christopher Newfield

"I'm really scared. I don't think we're going to make it."

These are the opening lines of a widely noted TED (Technology, Entertainment, Design—TED.com) talk delivered in March 2007. They were delivered not by an environmental activist but by the hardcore venture capitalist John Doerr. In the talk, Doerr traces his conversion experience to the culminating comment at a dinner conversation about global warming that was made by his 15-year-old daughter Mary. "I'm scared," she said. "And I'm angry. Dad, your generation created this problem. You better fix it" (Doerr, 2007).

Doerr's generation did not create global warming all by itself, but Doerr himself incarnates a technocratic vision of better living through ever-growing intensities of technology, and technology clearly caused the global warming problem. Doerr's talk continues this tradition. His solution to the problem of technology is not more technology, but green technology this time, made green through the efforts of market-based businesses and enterprise-oriented government policies. Doerr offered examples of changes made by companies, individuals, and governments that reduce both energy consumption and the production of greenhouse gases and invoked "radical innovation" as a secret ingredient, "innovating to zero," as Bill Gates put it in his TED Talk on the energy future three years later (Gates, 2010). Doerr, in effect, summarizes the linear model of innovation that drove the American "new economy" in the 1990s, in which governments provided basic research funding, research and development (R&D) tax breaks, a "High Protectionist" intellectual property climate (Goldstein, 2007), and some basic ground rules and then let companies and their profit expectations decide what products will be brought to market and with what environmental effects.

And yet each time Doerr mentioned one of these elements, he repeated his refrain that this actually will not be enough. And at the end of a final series of upbeat exhortations, Doerr found himself on the verge of tears. He was perhaps realizing that if he and the rest of us do not come up with much better innovations than those found in his talk, in 20 years, his daughter will not be speaking to him at all.

RADICAL INNOVATION FROM WHERE?

Where is such radical innovation to come from? One major source would appear to be nanotechnology. Over the past 10 years, nanotechnology has positioned itself as a leading emblem of radical innovation in a range of fields, energy included. The 2010 renewal campaign for the National Nanotechnology Initiative (NNI) featured energy challenges as a major focus of nanoscale concern (see Coe-Sullivan, 2010), and the NNI's Strategic Plan names, as one of three "signature initiatives," "[n]anotechnology for solar energy collection and conversion" (NNI, 2011). The NNI and the technology policy world look to basic and applied scientific research to solve the climate problem with substantive nanoscale breakthroughs that will lead to radical technological innovation.

But Doerr's—and our—failure to solve the climate problem is highly likely. The International Energy Agency (IEA), an authoritative analyst of global energy trends, defined the results of several scenarios in its 2010 World Energy Outlook. The status quo policy scenario is a complete climate disaster—irreversible warming with catastrophic economic and environmental results. The scenario in which CO_2 levels rise to "only" 450 parts per million will require policy changes that no governments are contemplating—and most climate scientists think 450 ppm is also too high. A middle Copenhagen-style scenario assumes that all pledges made at the climate summit are in fact implemented. Even if this unlikely event came to pass, CO_2 levels would climb to 650 ppm. Doerr certainly should still be scared—and his daughter should still be angry.

So to repeat the question, what would allow energy-related nanotechnological research to make a huge difference to this looming catastrophe? The most familiar answers are technological. There are many promising developments, most of them underway for years, and dozens of companies are trying to take advantage of them. Nanoscale energy technology is not my subject here, but there is little evidence of unusual speed or more frequent breakthroughs in energy-related nanoscale research than in other fields within the natural sciences and engineering.

The rate of acceleration required is formidable. The IEA (2010) finds that solar energy is on track to generate only 2 percent of world electricity by 2035—a shockingly low number that correlates with 650 ppm of CO_2. I extrapolate from IEA calculations that maintaining CO_2 levels at 450 ppm—already extremely high by historical standards—will require that the solar contribution to electricity generation increase by a factor of eight, or nearly an order of magnitude. Therefore, nanoscale research needs to help create a "Moore's law" in the efficient capture and use of energy such that an eight-fold increase would occur in about six years.

Another part of the answer to how to accelerate nanoscale research is increased research and development funding: It has been inadequate and is only gradually improving, but funding research is also not the topic of this chapter. Instead, my question in this chapter is, does nanotechnology as a *social* activity have some distinctive power that will allow it to accelerate innovation far more effectively than normal scientific practice?

ENTER COLLABORATION

Nanotechnology is an umbrella term that covers aspects of a wide range of scientific disciplines and research programs, nearly all of which have been underway for longer than the term has circulated. In seeking a special feature in the fields and practices that the term specifically designates, a leading candidate has for some time been thought to be *collaboration*.

Of course, collaboration does not mean a break with past scientific practice, but an extension of it. Price (1963) is often credited with noting that group collaboration had eclipsed solitary research and invention (Hackett, 2005), but by the 1950s, the success of large-scale collaborations such as the Manhattan Project had made the importance of both intensive and extensive collaboration clear to most practitioners. The development of two subsequent waves of technology reinforced collaboration's perceived centrality. Biotechnological advances are now widely viewed as depending on collaboration not just within one laboratory, but also among multiple laboratories that were often embedded in different kinds of institutions that include universities, government laboratories, and corporate research centers (Orsenigo, Pammolli, & Riccaboni, 2001).

Second, in the 1990s, information technology became as closely associated with the Internet as it had been with mainframe and then with workstations and personal computers, and the Internet appeared both to enable and to require a new extent and thoroughness of collaboration in scientific and all other domains. A large body of research has placed network collaboration at the center of modern science, and science's "webs of connection" are one of the central stories the contemporary social study of science (see Latour, 1987; Powell, White, Koput, & Owen-Smith, 2005). The same is true for the main line of analyses of commercialization and social adoption of innovations. These generally identify successful developments with complex forms of collaboration among heterogeneous institutions, including relationships between private and public institutions and the intricate, largely informal networks found in regional "clusters" (Hardagon, 2003; Robinson, Rip, & Mangematin, 2007; Rogers, 1995; Saxenien, 1994; Weil, Gallié, Mérindol, Lefebvre, & Pallez, 2010).

Science and technology studies (STS) has deepened our understanding of the continuous collaboration between technologists and society as a whole and has explored various practices via the concept of the co-generation of science and society (Jasanoff, 2004). In media theory and practice, the spread of Facebook, Twitter, LinkedIn, and similar services has encouraged many commentators to see social movements as the effect of social networks. Some Western commentators renamed the Green Movement opposing the official outcome of the Iranian election of June 2009 as the "Twitter revolution" (Harkin, 2010). Technology-mediated collaboration has become central to discussions of innovation in both society and science across a range of disciplines.

Given that most forms of interpersonal contact count, in most accounts, as some kind of meaningful collaboration, can we specify forms of nanoscale collaboration that have a clear positive impact on research?

Many observers have made the attempt. When nanotechnology became a subject of federal policy in the late 1990s, advocates proposed not only a wide range of interrelated projects but also augmented capacities for collaboration. Two of the five major categories funded by the NNI, which was launched in fiscal year 2001, involved building centers within networks and common infrastructure (Lane & Kalil, 2005). The Initiative promised augmented collaborations among government, university, and industry research centers, and subsequent NNI reports have also stressed the NNI's capacity to create interagency collaboration among the 25 federal entities that it has involved, as well as among different classes of universities, and the like in the United States and other countries. The FY 2009 NNI Budget Supplement had only a few pages that did *not* discuss collaboration in a wide variety of dimensions.[1]

The aspect of collaboration that has been important to discussions of nanotechnology has been interdisciplinarity. Nanotechnology has been seen as requiring the cooperation of people from many subdisciplines in engineering, biology, chemistry, physics, and many other fields. It has generated the most intense intellectual interest when it offers crossovers from one field to another, as suggested by attempts to use biological modes of self-assembly in electronics (see Huang, Notten, & Rasters, 2005). The NNI promised major benefits to the economy based in part on at least one major success in the commercialization of nanoscale research that had depended on bringing together groups that had not worked together before—in this case, metals and semiconductor researchers whose collaboration produced an important breakthrough in magnetic storage devices (McCray, 2009). The NNI considerably upped the ante for collaboration through the concept of *convergence*, championed by one of the initiative's principal architects, Mikhail Roco. In a series of papers (Roco, 2002, 2004, 2007; Roco & Bainbridge, 2001, 2003), Roco has argued for the ongoing convergence of formerly separate domains (nano-bio-info-cogno—or "NBIC" convergence). This convergence is eventually to transform the economy, the environment, and quality of life through a new "industrial revolution" (Roco, 2007).

Although many scientists have pointed out that nanotechnology is not a unified field, that it has highly diverse scientific roots, and that it may well not be the "general purpose technology" as was sometimes claimed (Lane & Kalil, 2005), there appears to have been a near-consensus in public discussions that 1) collaboration across disciplines and institutions is crucial to modern scientific progress, and 2) "nanotechnology," a complex ensemble of discourses, granting agencies (25), laboratories (60 research centers), companies (1,500), people (50,000), instruments, objects, and institutional networks, is encouraging and intensifying various forms of collaboration (see Roco, 2007).

Such claims have not been limited to policymakers and funding agency officials. Some scientists have also stated that the NNI has had a major impact on scientific collaboration in the United States:

> The real big impact of nanotechnology (and credit goes to NNI for this) on academia is the following—colleagues from different departments and disciplines have started talking to each other . . . NNI has almost forced this to happen . . . This is extremely positive, since the impact goes far beyond nanotechnology. This is a cultural change from Research [with]in Silos to Research without Borders. This scenario never would have occurred at Berkeley ten years ago, or even seven years ago. My enthusiasm in this manner of doing research . . . has influenced my students and my collaborators' students. If you now multiply this N times with all the other folks involved in NS&E [natural sciences and engineering], you can immediately realize the organic growth of influence of the NNI. (Teague, 2005)

The criterion for collaboration used by this scientist is similar to that found in analyses of Twitter revolutions: People in different domains "have started talking to each other." But *how* are they talking to each other? The range of practices here is enormous, and generalizations from any empirical set of activities are risky at best. Empirical data of sufficient detail would be ethnographic, and such data is relatively rare. Even if we did have this data, we would still need to determine what the effects of such discussions might be. Our question remains whether we can find a kind of collaboration in nanotechnological research that could at least partially justify the hope that it could transform energy research before global warming becomes fatally irreversible.

AMBIGUITIES OF COLLABORATION

Our own study of this question asked people what kind of collaboration they thought they were doing. Before discussing our findings, I will explain how this investigation of self-reported attitudes was motivated by research that studies the published evidence of collaborative practices.

The literature with which we engaged is largely bibliometric. In addition to having made important technical advances in the past decade, the bibliometric literature has also introduced two useful simplifications or stylizations of questions about the effects of scientific collaboration. The first is one already noted. Most of the literatures define the most interesting type of collaboration as that which crosses disciplinary lines. For example, an analysis may break out those electrical engineers who talk to marine biologists about cellular forms of self-assembly. It might focus on specialists in high-speed printing who are working on "nano-inks" that can carry electrical charges in ways required by next generation photovoltaic cells, and who have sought-out roofing contractors with experience with installing photovoltaic systems. Such interdisciplinary and cross-institutional discussions could combine technical, organizational, and social dimensions. As discussed subsequently, innovation is generally posited as heightened by crossing the boundaries of established disciplinary domains.

The second stylization is a distinction between information sharing among parties and joint knowledge creation. Although these activities cannot easily be told apart in practice, where they continuously overlap and combine, knowledge creation is generally seen as the more difficult and more powerful goal of collaboration. A study that helped us think about this distinction defines "deep collaboration" as research that is "jointly formulated," where each group "makes an essential contribution to different stages of the research process," and where there is interaction across a division of labor (Rafols & Meyer, 2007).

These two stylizations have helped develop the quantitative study of collaboration, and have underwritten the boom in bibliometrics, which studies quantities and patterns of citations and other features that link publications, as well as in network analysis of collaborative relationships (Rafols, Porter, & Leydesdorff, 2010). Many recent studies see the coauthorship of grant applications, articles, conference papers, and the like, as going beyond the basic exchange of information toward the kind of repeated and consistent interaction that suggests collaborative development of common ideas (Cressey, 2010).

And yet important ambiguities remain. For example, many studies correlate higher rates of connectivity among researchers with more innovation, using frequency of citation as their proxy of connectivity (Bornmann, de Moya Anegon, & Leydesdorff, 2010). Yet the fact that one cites an article in another field does not mean that one has a relationship either with the authors of the article or a positive absorption of the ideas and methodological traditions guiding the research on which the article reports. This is of course known to bibliometric researchers and is one of the reasons for the distinction between information exchange and knowledge creation, but this awareness does not change the fact that citation is best regarded as information exchange unless one knows something substantive about the relationships among the particular researchers in question.

The range of motives behind citation offers further reason to remain cautious in using co-citation and its variations as evidence of substantive collaboration. Citation can of course be critical or defensive. The latter motive is important in boundary-crossing work, where researchers may cite prominent authors and articles from their nonnative field to authenticate scholarly credentials that are not automatically recognized in the nonnative field. An economist working in network theory may cite leading network sociologists, for example, not because she agrees with the work, builds on it, or even grasps it fundamentally, but because she wants it as a credential or as currency in relation to her expanded audience that includes sociologists. When an article is already highly cited, non-informational uses may be more, rather than less, likely.

I am not suggesting that any of this is inappropriate, but, to the contrary, that it is normal. Motivation and comprehension are complicated, and published evidence must be interpreted carefully—ideally with individualized, qualitative empirical evidence.

Moving to group behavior, various studies suggest the ambiguous causality and unclear content in exchanges among research groups. For example, some research shows the absence of a straightforward relationship between scientific research and technological development in the same subdiscipline, even in the presence of citational links (Meyer, 2000, 2001). Huang et al. (2010) consider a body of studies and note, "No matter what conclusions they reach, the [cited] studies all use patent citations as a proxy for the linkage between science and technology," and at the same time, interviews of inventors behind ten selected nanotechnology patents found that "in only one of the ten cases one can draw a meaningful link between a patent and a particular publication that has stimulated the invention" (p. 150). In addition, a large proportion of knowledge in a technological area is tacit, and circulates through social and communicative networks in powerful ways that do not leave formal traces. Thus, even studies that assert that existing stocks of regional knowledge can be measured by patent counts understandably do not take on the far more difficult tasks of 1) identifying and characterizing specific knowledge pathways and 2) identifying the substantive information or knowledge that moves along those pathways (see Darby & Zucker, 2003; Rothaermel & Thursby, 2007; Zucker, Darby, Furner, Liu, & Ma, 2007). Finally, nanoscience seems to be subject to rates of patenting of early developments that have caused concern among scholars who are familiar with the extent to which weak intellectual property regimes have accelerated innovation in the past (Lemley, 2005; Mowery, 2011).

The situation is further complicated by ongoing methodological variation, including customized mixtures of citational and lexical search strategies: Huang et al. (2010) note that "almost every individual research group tends to develop its own search queries" (p. 157). I speculate that the continuing improvement of content analysis extraction tools, semantic tagging, and related techniques will proliferate a heterogeneous range of expensive reports, Internet journalism, informed amateur blogs, company publicity, and other types of usable but unverifiable and not clearly situated information. Knowledge pathways will likely become more diffuse, more complex, and in many cases more proprietary, self-interested, and, for the nonpaying scholar, incomplete. This would further complicate attempts to use formal publication, including patents, as proxies for information and knowledge exchange. In the midst of expansion and proliferation, formal publication may actually be in decline, as one study has recently confirmed after years of delaying publication of their suspected findings (Leydesdorff & Meyer, 2010). We can conclude that publication and patent counts and network maps are, in the language of textual theory, *symbols* rather than literal *signs* of scientific or technological advancement. They require careful, case-by-case interpretation (Newfield, 2010a, 2010b).

In short, the relation between collaboration and innovation is not at all obvious, and the same is true for both sides of the equation. Certainly the literature in several disciplines supports the idea that crossing intellectual boundaries is a key element in the creation of new ideas (Amabile, 1996; Csikszentmihalyi, 1996; Simonton, 2004), and "a series of 'translations' from one context to another is the mill from which new 'ideas' are generated and pursued in the course of laboratory research" (Doing, 2008). Nonetheless, these lingering ambiguities make it hard to identify a collaborative power that is specific to nanotechnological science and technology in its myriad forms (NSTC, 2011).

My research group sought to start with something very simple. Is there a correlation between saying that you work in nanotechnology and more frequent collaborating *across* disciplines? If there is quantitatively more of this crossing, then perhaps we could move on to speculate about *qualitative* changes in science's social life as wrought by nano.

DO YOU COLLABORATE?

Because scientific research is already so collaborative, nanotechnology may be a straightforward continuation of existing trends and resemble existing fields—particularly "big science" fields like high-energy physics—that rely on complex combinations of skills and expensive and/or relatively scarce equipment (Hackett, Conz, Parker, Bashford, & DeLay, 2004). However, nanotechnology as an umbrella "field" has been so consistently defined as interdisciplinary that we should expect higher levels of interdisciplinarity conviction and practice. We conducted a pilot survey that asked a range of questions about the participants' affective relation to collaboration (see the final section of this chapter, "Methods").

Our survey asked respondents about the extent to which they considered their research to involve nanotechnology. We also asked them a series of questions about the extent and nature of their interdisciplinary contacts. Our discussions and interviews that led up to the survey suggested that a portion of identification with nanoscale research was opportunistic. Some of our subjects said that their colleagues had become "nanotechnologists" because the NNI provided a new source of research funds. In addition, as already noted, the term *nanotechnology* covered research domains that in most cases had been in development for years or decades under others names (McCray, 2009). The term has not achieved meaningful currency in the broader culture (Weaver & Bimber, 2008) such that researchers would be able to use it to explain who they were or what they did. But because we regarded nanotechnology as a potential *professional* identity and because professional identities can be fluid and negotiated (Beaulieu, Rioux, Rocher, Samson, & Boucher, 2008; Callan et al., 2007) and because "singular and multiple [identities] can each be called on [and] can work together" to "maintain effectiveness in different environments" (Griffin, 2008), we felt that it would still be useful to measure self-identification with nanotechnology, acknowledging permeability between research practice and emerging professional identity (Lamb & Davidson, 2005).

We then asked whether researchers who identify themselves as nanoscale researchers collaborate more frequently across disciplinary lines. Our questions explored two dimensions of this issue. First, do nanotechnologists engage in more interdisciplinary collaboration than do other scientists in the sense of being more likely to work with more than one laboratory at a time? Second, do nanoscale researchers *want* to engage in interdisciplinary collaboration, or do they do it because it is expected of them, as a basic part of their job?

Nearly 50 percent of our sample said that their work involved *no* nanoscale research at all. At the other extreme, about 5 percent of our sample reported that all of their work was at the nanoscale. The remaining 45 percent of the sample was distributed at low rates among varying percentages, with some lumping at 20 percent and 50 percent. For purposes of the correlations we describe next, we organize our sample into three groups: "no nano," whose members said they performed no nanoscale research; "some nano," which included everyone from 1 percent to 50 percent; and "majority nano," which referred to researchers who felt that more than 50 percent of their research was conducted at the nanoscale in relation to size-dependent phenomena.

In response to questions about whether they considered themselves to be nanotechnologists, only 1 percent said they strongly agreed with the statement that they "generally" consider themselves to be nanotechnologists. Whereas about 20 percent of participants said their laboratory work was "mostly nano," only 15 percent felt any degree of agreement, including "somewhat agree," with a statement asking whether "nanotechnologist" was part of their professional identity. A further question also produced an interesting result. Among the subset of subjects who said that the majority of their work was at the nanoscale, nearly as many nanoscale researchers strongly denied the nano label as strongly affirmed it. In our sample, the professional identity of either nanoscientist or nanotechnologist was not crystallizing even among those who do most of their research at the nanoscale.

What about collaboration across interdisciplinary lines? We correlated nanoscale identity and nanoscale laboratory activity with the number of associations to labs outside the primary laboratory. We found that about half of non-nano researchers have affiliations with at least one other laboratory. This finding fits with the general view that scientific disciplines are all shifting toward inter-institutionality and interdisciplinarity. But the shift is stronger at the nanoscale. When both dimensions are combined, 90 percent of the small group whose work is majority nano *and* who is strongly identified as nanoscientists or nanotechnologists has multi-laboratory affiliations. This pilot study suggests that nanotechnologists *do* engage in more interdisciplinary collaboration than do other researchers.

NEW CAPABILITIES FROM COLLABORATION?

The larger motivation for our study was to discover whether cross-disciplinary collaboration was creating unusual intellectual chemistry, a heightened capacity for innovation, or strange events that might lead somewhere unexpected. We did not want to ask about this directly, given both the ambiguity in the terminology and the likelihood of biasing responses. We instead asked a series of questions about experiences of collaboration—its practicality, pleasures, and usefulness, among other things.

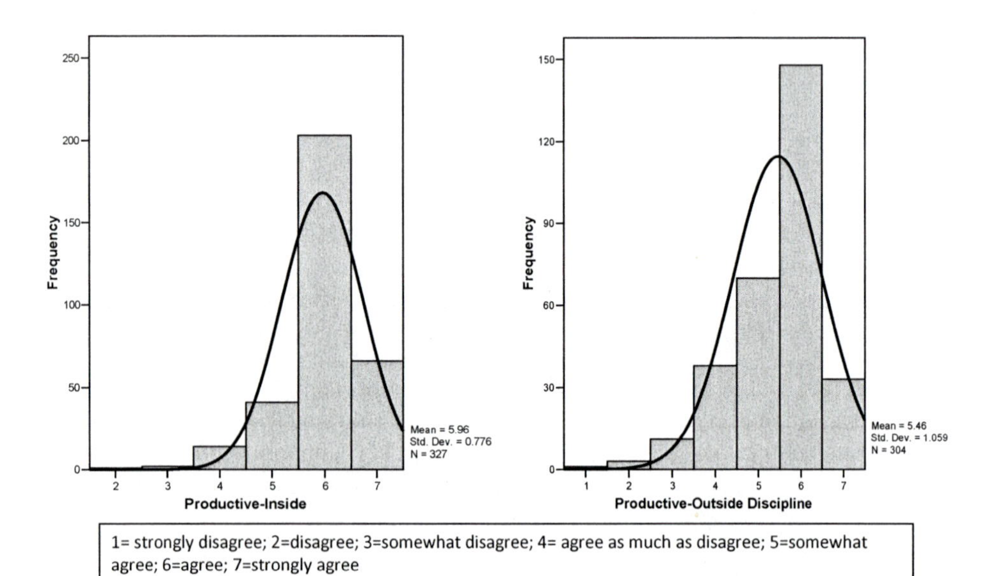

Figure 5.1 We asked whether "collaborating with scientists inside my discipline" is "productive" and asked the same question about collaboration *outside* the discipline.

Large majorities of respondents believed that collaboration was generally productive, indicating an overall consensus that has long blossomed in the sciences. But as a group they felt that collaboration *within* the discipline was more productive than collaboration outside of it: About 95 percent somewhat agreed, agreed, or strongly agreed that it was productive inside the discipline, whereas 82 percent thought collaboration was productive outside the discipline. Among the true partisans of collaboration, those who "strongly agreed" with its value, about twice as many valued it inside rather than outside the discipline (20.2 percent compared to 10.9 percent).

When we asked further questions about the motivations behind collaboration outside of one's discipline and then sorted the data into our categories of no nano, some nano, and majority nano, we found that majority nano researchers were only slightly more likely to seek opportunities to collaborate outside the discipline than were other researchers. Goals were familiar: Forming professional connections was the most commonly mentioned motivation, with the chance for increased publications and research funding not far behind. Respondents did not avail themselves of the opportunity to write-in other motives for collaboration: intellectual development, improved knowledge-creation, advancement of science, solving major scientific problems, and others. The study did not find the early signs of a distinctive outlook or distinctive culture of collaboration among nanoscale researchers. We concluded as follows: Although collaboration is seen as productive by large majorities of researchers, it appears to function *more as a required baseline of disciplinary scientific practice than as a productivity-enhancing activity.* And collaboration outside the discipline was seen as less productive.

The following general picture thus appears to be the most likely. Inter-institutional collaboration has indeed become common in scientific practice, and nanoscience is no exception. Policy discourse has developed a regulative ideal for nanoscale practice, which ideal is a perpetual openness to collaboration across institutional and disciplinary boundaries. Researchers who work at the nanoscale do in fact tend to have more interdisciplinary collaborations in the specific sense we have defined. At the same time, nanoscale practice has *not* yet come to support higher levels of positive interest in collaboration outside the home discipline. Although the policy discourse of nano-enabled collaboration is powerful and often inspiring, it has not transformed the attitudes or the experiences of rank-and-file nanoscale researchers like those captured by our sample. In other words, in terms of its internal social relations, the social life of nanotechnology is much like the social life of the rest of science.

Within the limits of this pilot survey, whose results cannot be regarded as decisive, it does not seem that any novel feature about nanoscale collaboration as such will allow nanotechnology to accelerate the development of solutions of a major challenge like the growing climate emergency—without other systemic changes. In other words, the broader social effectiveness of nanoscale science and technology will likely not depend not anything intrinsic to nanotechnology, but, as has historically been the case, on science and social policy overall.

NORMAL NANOTECHNOLOGY

These results do suggest a series of steps that might strengthen support for nanotechnological research while enhancing its actual results. Here, I move beyond the indications of the survey instrument itself and base my remarks on my immersion in a range of interviews, discussions with participations, and observation of legal and policy discussions since 2004.

The first of these is a simple framing shift. Rather than seeing nanoscience as extraordinary science in Thomas Kuhn's sense, we should see it as *normal science*. Similarly, we should see nanotechnology as normal technology, having an impact on society over time, through complex patterns in which technologists and policymakers are in continuous touch with end-users through democratized feedback mechanisms (von Hippel, 2005).

Some interesting consequences follow once we bite the normalcy bullet. The first, which is simple but crucial, is that we can no longer assume that nanotechnology will progress more rapidly than ecological biology or drug development or anything else. We should instead assume that it will progress at a normal pace, given its normal forms of collaboration.

This view does not rule out major breakthroughs or periods of amazingly rapid growth. It does say that we can look to the recent history of science to estimate the pace at which developments will occur. For example, the applications for the first of the Cohen-Boyer patents that helped found the biotechnology industry were filed in November 1974. In spite of massive investment, a global array of companies, and claims of nearly 3,000 products, industry profitability has never quite been achieved, and the most transformational medical products have not yet arrived (see Timmerman, 2011). A maturation span of 50 years is not, in fact, abnormal. Perhaps we should assume the same half-century developmental frame for nanotechnology—under our current innovation model.

The second consequence of nanoscale's normal collaborative practices and normal advantages is that policymakers, university administrators, and the general public need to be *told* that, with the current system, nano-progress will be coming slowly. Politicians, executives, voters, schoolchildren—everybody needs to be told about the pace at which knowledge and technology generally move forward. Great expectations can be maintained, but only with a realistic timeframe that will not cause disappointment, frustration, cynicism, drifting attention, backlash, and other staples of science policy.

A further effect of nano-normalcy is that, in the absence of *intrinsic* powers of accelerated development, nanoscience and technology's various disciplines may attract increased financial investment. For its first 10 years, some NNI promoters implied that four generations of nanoscale development would take place for something like $1 billion per year from the federal government, corrected over inflation, or say $30 billion of public investment over 20 years. This is a short time and a preposterously small figure to spend on a wide range of technologies: The highly focused Manhattan Project spent over $20 billion in today's dollars in less than four years on just one product, the atomic bomb. If NST advocates state clearly and repeatedly that it is advancing like every other field, neither more quickly nor more slowly than theirs, it will, ironically, be eligible for larger amounts of basic funding.

A fourth consequence: If we do not yet actually know how super-effective collaboration might take place in NST, then we are in a position to work hard to find out. We should put collaborative needs on the social agenda. We will be able to ask openly the fundamental questions: What is missing from nanoscale collaboration as currently practiced? What do practitioners need to do their work better? How can we provide the institutional conditions that will make this happen?

A further consideration is the identity of the "we" that would conduct this inquiry. In fact, "they" must be representatives of nano's social life. This means not only the upstream policymakers, agency managers, and principal investigators who have shaped the nanoscale agenda since its inception, but also the downstream members of society who have a stake in the precise effects of nanoscale research—the problems being tackled, the products being envisioned, the outcomes desired not only by practitioners but by society itself (see Cozzens, Gatchair, Kim, Ordóñez, & Supnithadnaporn, 2008). This set of social procedures would develop nanoscale research in response to systematically determined popular objectives and popular objectives. In a process often implied but never implemented, experts would not decide and deliver products to the public. Instead, the public would decide, and the experts would figure out how to refine and deliver what public deliberation had projected.

Even an outline of these procedures is beyond the scope of this paper (Newfield & Boudreaux, forthcoming), but a few conditions can be mentioned here.

They cannot occur *only* at the microscale, and infer a whole direction from consultation of a tiny part. Focus groups are frequently used to sound "public opinion," but these are microcosms rather than social processes. The full-scale investigation of nano-collaboration will clearly involve the Internet, and use what we know about two apparently opposed methodologies: face to face conversation, and massively multi-user environments that enable crowd sourcing and other large-scale practices of consultation.

The participating groups must be selected in an egalitarian way: nontechnical people, teachers, stonemasons, accoutants, bass players, chambermaids, parking lot attendants—everyone who is being affected by the issues the research. Studying nanoscale collaboration's full social meaning will require overcoming the standard divisions of both technical and political labor that maintain intellectual segregation and the non-communication of ideas.

The same must be said of the integration of modes of thinking within a particular group or professional discipline. Universities have always been divided into disciplines, and later into departments, divisions, and schools much like the multidivisional corporations with which they sometimes cooperate on research. But the original genius of the university was to offer a secure and supportive venue for *combining reflection with reflection upon the process of reflection.* This is one of the primal activities of philosophy as such, which seeks to understand the processes of reason and the various forms they take. Knowledge is a kind of self-activity, in the terms of the German philosophers, particularly Humboldt and Fichte, who helped create the university as we know it. This means that it results not only from a set of research procedures but from ongoing processes of analysis of those procedures, which should be part of the research activity itself.

What this means is that nanoscale scientists and engineers should not farm out social analysis to professional social scientists and humanists, but must engage in it as part of their firsthand experience with the nanoscale work as such. They cannot do it alone, and would need collaboration with and guidance from social scientists and humanists, but would need to remain active in a social engagement with their own scientific practice—with its institutional framework, its funding, its effects on its participants including students and staff and non-participating units in its own institutions such as universities, its immediate expectations, its larger worldly hopes. Scientists can and should share these inquiries with social and cultural specialists in interdisciplinary teams. The main point is that the research and the social life of the research must be copresent and developed by the same people, not, as is currently the case, split apart.

Finally, the same point must be made about the relationship between society and business. John Doerr, the venture capital community, and the leaders of dominant regional clusters like Silicon Valley have systematically advanced two forms of privatization. One is the privatization of the financial returns of technology development, even where they have enjoyed generous public funding. The other is the privatization of the product development process, in which small groups of executives working in closely-knit networks make all major decisions about product development in near-total and routinized secrecy. The constant presence in these private deliberations of the future market value of a product is generally assumed to provide sufficient public accountability, because a no vote from consumers can kill any product. But market calculations cannot substitute for popularly expressed social goals, and this is why public goods are subject to "market failure," as innovation economists know perfectly well.

Opening the social processes is more urgent than ever given failure of Silicon Valley to make major contributions to alternative energy development in the years since Doerr's 2007 speech. This problem is particularly acute given that photovoltaic modules have used the same silicon processing methods in which it has world-leading expertise. What we have seen instead of Silicon Valley breakthroughs is a commoditization of existing technology by low-cost producers, particularly from East Asia, that has pushed financing away from next-generation products (see Woody, 2011). In short, it is quite likely that if John Doerr wants to stop being scared about the future climate, he needs to help to open the technology decision process to the wider society briefly described above. The business sector needs, in short, to allow market-based development to give way to nanotechnology's larger social life.

METHODS

Survey data were collected in the fall of 2007 at a large university in the Western United States with a reputation for pioneering nanoscale research and possessing several major nano-specific research organizations. The sample was purposive; a roster was first obtained that had the names of all students, faculty, postdoctoral employees, and others who had received monetary compensation for laboratory work at this university. The total came to 1,939 individuals. Attempts were made to insure that the full range of science and engineering disciplines were represented, as we wished to include researchers from outside of nanotechnology. Subjects who inquired as to the appropriateness of their participation (from statistics or marine biology, for example) were almost always encouraged to participate. We anticipated that a large majority of respondents would be graduate students. We were very interested in the views of younger researchers who were receiving their graduate educations after the NNI had been in place for some years, and when interdisciplinarity had become a common expectation.

Participants were notified via e-mail about the opportunity to answer a survey regarding innovation and collaboration in the laboratory. An embedded link in the e-mail assigned the participant a unique identifier number to ensure privacy and confidentiality. The confidentiality waiver stressed that no names would be used in the report of data. In addition, participants remained anonymous to the researchers.

After online consent was given, the participant took an online survey with six sections: demographics, laboratory characteristics, extent and type of interorganizational collaboration, descriptions of research domains, communication assessments, and attitudes toward innovation. A high ratio of graduate students to faculty was expected, and participants were offered $10 compensation at a later date. After completing the survey, participants were directed to an office on campus to receive payment.

The original response rate was about was about 38.6 percent (749 responses), but this figure does not reflect the final sample, which was cleaned to eliminate duplications, undergraduates, several categories of departed staff, and very incomplete or spoiled responses. The final participant pool ($N = 420$) captured a broad disciplinary breakdown ($N = 399$).

After the demographic questions, participants were asked to describe their research and their research funding. They then answered laboratory identification questions (Mael & Ashford, 1992; Obst & White, 2005). Participants answered a series of laboratory communication climate questions (Ekvall, 1986), and responded to a five-tree network structure for collaborations with other labs. The interorganizational collaboration section of the survey contained most of the dependent measures. Participants were asked to self-report their publications, and to discern what percentage of these publications involved collaborations across disciplines, among different institutions, and in relation to nanotechnology. This section also asked about basic attitudes toward collaboration, and toward information-seeking skills (Borgatti & Cross, 2003). The fourth area of the survey attempted to ascertain perceived boundaries of the nanotechnology domain (NNI, 2007; Shapira, 2006) and asked a series of questions that dug more deeply into the possibility that nanotechnology was becoming a meaningful professional identity, including questions about identification with nanotechnology, the prestige of nanotechnology as a new discipline, interpersonal affects in relation to nanotechnologists, the perceived benefits of nanotechnology, and a restructured group permeability (Mael & Ashforth, 1992; Obst & White, 2005; Tajfel & Turner, 1979). The fifth area of the questionnaire asked communication-related questions, mainly items regarding communication apprehension (McCroskey, PRSA-24). Other questions were taken from the individual innovativeness scale (Hurt, Joseph, & Cook, 1977), the willingness to collaborate scale (Anderson, Martin, & Infante, 1993), and a measure of writing apprehension (Daly, 1975). The final section involved derived questions about whether publishing, patenting, and/or commercialization were important motives in guiding nanotechnological research and innovation.

ACKNOWLEDGMENTS

This material is based upon work supported by the National Science Foundation under Cooperative Agreements Nos. SES 0531184 and SES 0938099 to the Center for Nanotechnology in Society at University of California at Santa Barbara. Any opinions, findings, and conclusions or recommendations expressed in this material are those of the author and do not necessarily reflect the views of the National Science Foundation.

NOTES

1. There were thirty-four occurrences of the root term in thirty-seven pages of text (see NNI, 2009).

REFERENCES

Amabile, T. M. (1996). *Creativity in context: Updates to the social psychology of creativity*. Boulder, CO: Westview.

Beaulieu, M. D, Rioux, M., Rocher, G., Samson, L., & Boucher, L. (2008). Family practice: Professional identity in transition. A case study of family medicine in Canada. *Social Science and Medicine, 67*(7), 1153–1163.

Bornmann, L., de Moya Anegón, F., & Leydesdorff, L. (2010). Do scientific advancements lean on the shoulders of giants? A bibliometric investigation of the Ortega hypothesis. *PLoS ONE, 5*(10), e13327. doi:10.1371/journal.pone.0013327. Retrieved on 6/4/11 from www.plosone.org/article/info:doi/10.1371/journal.pone.0013327

Callan, V. J., Gallois, C., Mayhew, M. G., & Grice, T. A., Tluchowska, M., & Boyce, R. (2007). Restructuring the multi-professional organization: Professional identity and adjustment to change in a public hospital. *Journal of Health and Human Services Administration, Spring*, 448–477.

Coe-Sullivan, S. (2010, December). QD vision's quantum light optics: Delivering on the promise of nano. Presentation to NNI Innovation Summit Program. Retrieved on 4/13/11 from http://www.nsti.org/events/NNI/program/

Cozzens, S. E., Gatchair, S., Kim, K., Ordóñez, G., & Supnithadnaporn, A. (2008). Knowledge and development. In E. J. Hacket, O. Amsterdamska, M. Lynch, & J. Wajcman (Eds.), *The handbook of science and technology studies* (3rd ed.), pp. 787–813. Cambridge, MA: MIT Press.

Cressey, D. (2010, October 13). Counting collaboration. *Nature News*. Retrieved on 3/14/11from http://www.nature.com/news/2010/101013/full/news.2010.538.html

Csikszentmihalyi, M. (1996). *Creativity: The flow and psychology of discovery and invention*. New York: Harper Collins Publishers.

Darby, M. R., & Zucker, L. G. (2003). *Grilichesian breakthroughs: Inventions of methods of inventing and firm entry in nanotechnology* (National Bureau of Economic Research Working Paper Series No. 9825). Retrieved on 9/13/11 from http://www.nber.org/papers/w9825

Doerr, J. (2007, March). John Doerr sees salvation and profit in greentech. *Ted Talks*. Retrieved on 6/5/11 from http://www.ted.com/talks/john_doerr_sees_salvation_and_profit_in_greentech.html

Doing, P. (2008). Give me a laboratory and I will raise a discipline: The past, present, and future politics of laboratory studies in STS. In E. J. Hackett, O. Amsterdamska, M. Lynch, & J. Wajcman (Eds.), *The handbook of science and technology studies* (3rd ed.), pp. 279–297. Cambridge, MA: MIT Press.

Gates, B. (2010, February). Innovating to zero! *Ted Talks*. Retrieved on 6/5/11 from http://www.ted.com/talks/lang/eng/bill_gates.html

Goldstein, P. (2007). *Intellectual property: The tough new realities that could make or break your business*. London: Penguin/Portfolio Hardcover.

Griffin, A. (2008). "Designer doctors": Professional identity and a portfolio career as a general practice educator. *Education for Primary Care, 19*, 355–359.

Hackett, E. J. (2005). Introduction to the special guest-edited issue on scientific collaboration. *Social Studies of Science, 35*(5), 667–671.

Hackett, E. J., Conz, D., Parker, J, Bashford, J, & DeLay, S. (2004). Tokamaks and turbulence: Research ensembles, policy and technoscientific work. *Research Policy, 33*(5), 747–767.

Harkin, J. (2010, 2 December). Cyber-con. *London Review of Books, 32*(23), 19–21.

Huang, C., Notten, A., & Rasters, N. (2010). Nanoscience and technology publications and patents: A review of social science studies and search strategies. *Journal of Technology Transfer, 36*(2), 145–172.

Huang, Y., Chiang, C., Lee, S. K., Gao, Y., Hu, E. L., De Yoreo, J. & Belcher, A. M. (2005). Programmable assembly of nanoarchitectures using genetically engineered viruses. *Nano Letters, 5*(7), 1429–1434.

Hurt, T. H., Joseph, K. & Cook, C. D. (1977). Scales for the measurement of innovativeness. *Human Communication Research, 4*(1), 58–65.

International Energy Agency (IEA). (2010). *World energy outlook, 2010.* Retrieved on 3/26/11 from http://www.worldenergyoutlook.org/

Jasanoff, S. (2004). *States of knowledge: The co-production of science and the social order.* London: Routledge.

Lamb, R., & Davidson, E. (2005). Information and communication technology challenges to scientific professional identity. *The Information Society, 21,* 1–24.

Lane, N. & Kalil, T. (2005). The National Nanotechnology Initiative: Present at the creation. *Issues in Science and Technology, Summer,* 49–54.

Latour, B. (1987) *Science in action: How to follow scientists and engineers through society.* Cambridge, MA: Harvard University Press.

Lemley, M. A. (2005). Patenting nanotechnology. *Stanford Law Review, 58,* 601–630.

Leydesdorff, L., & M. Meyer. (2010). The decline of university patenting and the end of the Bayh-Dole effect. *Scientometrics 83*(2), 355–362

Mael, F., & Ashforth, B. E. (1992). Alumni and their alma mater: A partial test of the reformulated model of organizational identification. *Journal of Organizational Behavior, 13*(2), 103–123.

McCray, W. P. (2009). From lab to iPod: A story of discovery and commercialization in the post–Cold War era. *Technology and Culture, 50*(1), 58–81.

Meyer, M. (2000). Patent citations in a novel field of technology—what can they tell about interactions between emerging communities of science and technology? *Scientometrics, 42*(2), 151–178.

Meyer, M. (2001). Patent citation analysis in a novel field of technology: An exploration of nano-science and nano-technology. *Scientometrics, 51*(1), 163–183.

Mowery, D. C. (2011). Nanotechnology and the U.S. national innovation system: Continuity and change. *Journal of Technology Transfer, 36*(6), 697–711.

National Nanotechnology Initiative (NNI). (2009). *National Nanotechnology Initiative annual report, fiscal year 2009.* Retrieved on 7/15/10 from http://www.nano.gov/NNI_09Budget.pdf

National Nanotechnology Initiative (NNI). (2011). *NSTC 2011.* Retrieved from http://www.nano.gov/sites/default/files/pub_resource/2011_strategic_plan.pdf

Newfield, C. (2010a). Avoiding network failure: The case of the National Nanotechnology Initiative. In F. Block & M. Keller (Eds.), *State of innovation: The U.S. government's role in technology development.* New York: Paradigm Press, pp. 282–99.

Newfield, C. (2010b, December 20). Science out of the shadows: Public nanotechnology and social welfare. *Occasion, 2.* Retrieved on 9/13/11 from http://occasion.stanford.edu/node/61

Newfield, C., & Boudreaux, D. (forthcoming). Introduction. In C. Newfield & D. Boudreaux (Eds.), *Can rich countries still innovate?*

Obst, P., & White, K. (2005). Three-dimensional strength of identification across group memberships: A confirmatory factor analysis. *Self and Identity, 4*(1), 69–80.

Orsenigo, L., Pammolli, F., & Riccaboni, M. (2001). Technological change and network dynamics: Lessons from the pharmaceutical industry. *Research Policy, 30,* 485–508.

Powell, W. W., White, D. R., Koput, K. W., & Owen-Smith, J. (2005). Network dynamics and field evolution: The growth of interorganizational collaboration in the life sciences. *American Journal of Sociology, 110*(4), 1132–1205.

Price, D. J. de Solla. (1963). *Little science, big science*. New York: Columbia University Press.

Rafols, I., & Meyer, M. (2007). How cross-disciplinary is bionanotechnology? Explorations in the specialty of molecular motors. *Scientometrics, 70*(3), 633–650.

Rafols, I., Porter, A. L., & Leydesdorff, L. (2010). *Science overlay maps: A new tool for research policy and library management*. Retrieved December 14, 2011, from http://www.leydesdorff.net/overlaytoolkit/overlaytoolkit.htm

Robinson, D. K. R., Rip, A., & Mangematin, V. (2007, July). Technological agglomeration and the emergence of clusters and networks in nanotechnology. *Research Policy, 36*(6), 871–879.

Roco, M. C. (2002). Coherence and divergence in science and engineering megatrends. *Journal of Nanoparticle Research, 4*(1–2), 9–19.

Roco, M. C. (2003). Converging science and technology at the nanoscale: Opportunities for education and training. *Nature Biotechnology, 21*(10), 1247–1249.

Roco, M. C. (2004). Nanoscale science and engineering: Unifying and transforming tools. *AIChE Journal, 50*(5), 890–897.

Roco, M. C. (2007). National Nanotechnology Initiative: Past, present, future. *Handbook on nanoscience, engineering and technology* (2nd ed.), pp. 3.1–3.26. Boca Raton, FL: Taylor and Francis.

Roco, M. C., & Bainbridge, W. S. (Eds.). (2001). *Societal implications of nanoscience and nanotechnology*. Boston: Springer.

Roco, M. C., & Bainbridge, W. S. (Eds.). (2003). *Converging technologies for improving human performance*. Boston: Springer

Rogers, E. M. (1995). *Diffusion of innovations* (4th ed.). New York: The Free Press.

Rothaermel, F. T., & Thursby, M. (2007). The nanotech versus the biotech revolution: Sources of productivity in incumbent firm research. *Research Policy, 36*(6), 832–849.

Saxenien, A. (1994). *Regional advantage: Culture and competition in Silicon Valley and Route 128*. Cambridge, MA: Harvard University Press.

Simonton, D. K. (2004). *Creativity in science: Chance, logic, genius, and zeitgeist*. New York: Cambridge University Press.

Teague, E. C. (2005). *NNI contributions to nanotechnology commercialization*. Retrieved on 9/13/11 from http://www.nsti.org/NanoImpact2005/speakers/NNI_NIS.pdf

Timmerman, L. (2011, May 16). Will biotech ever again captivate the public imagination, Like Facebook or LinkedIn? *Xconomy*. Retrieved on 5/27/11 from http://www.xconomy.com/national/2011/05/16/will-biotech-ever-again-captivate-the-public-imagination-like-facebook-or-linkedin/

Von Hippel, E. (2005). *Democratizing innovation*. Cambridge, MA: MIT Press.

Weaver, D. A., & Bimber, B. (2008). Finding news stories: A comparison of searches using LexisNexis and Google News. *Journalism and Mass Communication Quarterly 85*(3), 515–530.

Weil, T., Gallié, E-P., Mérindol, V., Lefebvre, P, & Pallez, F. (2010, June) *Why are good comparative studies of networks so rare? Practical lessons from a study on French clusters* (CERNA Working Paper Series, Working Paper 2010–08, Ecole des Mines). Retrieved on 9/11/13 from http://hal.archives-ouvertes.fr/docs/00/48/84/04/PDF/CWP_2010–08.pdf

Woody, T. (2011, August 31). What Solyndra's bankruptcy means for Silicon Valley solar startups. *Forbes*. Retrieved on 10/12/11 from http://www.forbes.com/sites/toddwoody/2011/08/31/what-solyndras-bankruptcy-means-for-silicon-valley-solar-startups/.

Zucker, L. G., Darby, M. R., Furner, J., Liu, R. C., & Ma, H. (2007). Minerva unbound: Knowledge stocks, knowledge flows and new knowledge production. *Research Policy, 36*(6), 850–863.

6 Working for Next to Nothing
Labor in the Global Nanoscientific Community

Mikael Johansson

Ethnographic fieldwork can sometimes be tedious work. Day after day you collect massive amounts of data, data for which you do not see any imminent use. My first fieldwork among nanoscientists was in 2003–2004 at the Department of Microtechnology and Nanoscience at Chalmers University of Technology, located in Gothenburg, Sweden. The scientists I followed experimented on the nanometer level, the level of single atoms and molecules. Nanoscience is also often referred to as the "next big thing" in research; compared to other disciplines, it is well funded, and there are high expectations about possible outcomes (Berube, 2006; Lok, 2010). The high level of interest in nanoscience and nanotechnology is partly explained by the significant expected economic outcomes but also by the more metaphysical notion of experimenting with nature's "next to nothing," atoms and molecules. Atoms and molecules are the last stable building blocks in our physical world and hence represent a "final frontier" for technoscientific advances.

In the days of my first fieldwork, I was entirely focused on completing my PhD, and the aim of the research was the cosmology of scientists. I built on this research during a subsequent postdoctoral research position in Santa Barbara in 2009–2010 by adding a second year of laboratory fieldwork among not only United States–based nanoscientist's but also among toxicologists studying the potential adverse effects of nanoparticles. During my two years at the University of California at Santa Barbara I also became the president of the postdoc association and volunteered for the fledging postdoc union. It was during this time I started to think about labor issues in relation to higher education. Looking back through my field notes, from both Sweden and the United States, I discovered many bits and pieces about scientific work and the scientific workplace that had gone unnoticed before.

The question that interested me and became the focal point for this text is, why do nanoscientists and nanoscientists-in-training accept what I would call poor working conditions? In interviews, people described how they worked 60–90 hours a week, had little or no vacation time, working evenings, nights, weekends, and many of them with a surprisingly poor salary. The scientists I talked to are not your low-income, low-educated, working-class group, fighting for survival often without a voice in society (see Bourgois, 2003). When asked about their social background, it became pretty clear a vast majority of the scientists come from middle-class backgrounds (Johansson, 2009). The middle class, in a broad sense, is composed of those occupations and associated lifestyles that are reached by merit through education or other qualifications, giving its holders greater material and social benefits than those acquired by manual labor, that is, the working class (Giddens, 2001). In Sweden, for example, it is estimated that 75 percent of children from academic families continue with higher studies, whereas the working class is grossly underrepresented (Frykman, 1986). The middle class has also in a sense become the norm of Western societies (see Lawrence, 2006; Löfgren, 1987) and thus, I found it surprising that the group of middle-class scientists do not protest more about their work situation.

After some pondering and asking around, I found three important factors that regulate why nanoscientists do not complain about their working conditions. The first factor is the dependence on social networks. To have a scientific career the individual scientist is dependent on their peers. To complain about workloads and working hours can be seen as troublemaking and may contribute to the view that the individual scientists do not pull their load in the lab. In a research group, individual careers are intertwined, and if one individual starts "slacking," it will have consequences for the others. This type of dependency on one's peers is a bit different from similar, more hierarchal types of social labor control. It is not a question of fear from displeasing one's boss (see McKay, 2006); it is a fear of displeasing one's boss and colleagues, people with similar status within the lab. Second, there is a scientific culture of endurance. People who complain are seen as not being cut out to be scientists. The norm is to work weekends and nights, and those who do not can be seen as weaklings. The third factor is a belief among many of the scientists I talked to that their hard work period is a transitory phase. It is seen as perfectly fine to put in extra hours when you are young and building your career, and later in life, the scientists can work less, a notion that seems to be more of an ideal then reality.

The three factors—dependency on social networks, culture of endurance, and hard work as a transitory phase—have to be framed in a grander picture. The nanoscientific community is hierarchal with professors at the top as gatekeepers. They decide who will be let into the community, and they are key players in transferring cultural values to the younger generation. To accumulate scientific credibility within the community, there is a need to publish lots of scientific articles, something that is done through collaboration with colleagues. It is accordingly important for each individual to work hard, not only for their own career, but also for their colleagues' career. This is something that leads to a hectic work situation with a cultural norm of hard work and no complaint. According to the cultural norm, those who do complain are just not cut out to be nanoscientists! In a sense, the work life of the nanoscientist, with hardship, sacrifice, and potential reward can be simile to the universal myth of the archetypical hero (Campell, 2008). There is also a scientific ethos of self-invisibility (Haraway, 1997) in which the scientists should be objective and not place themselves in the center of attention. To complain about workloads, for example, increases their subjective responses to the scientific process, something that should be avoided.

Figure 6.1 Ordinary sunlight can interfere with experiments on the nanometer scale. Thus, in many labs, the scientists work without seeing sunlight during the day. Photo has been edited to protect the anonymity of the researchers.

The nanoscientific community is a male-dominated community. At the nanolaboratory in Sweden, approximately 90 percent of the workers were men (Johansson, 2009). When going through the ethnographic material for this paper, however, I realized that females are overrepresented in my interviews dealing with labor issues. A reason for this might be that females, being a minority and more often primary family care takers, both in Sweden and the United States, reflect more on these issues than their male counterparts. Minority groups often have better insights of community norms than those who act within the majority of the norm (Abu-Lughod, 1991; Keesing, 1994).

SCIENTIFIC LABORERS

In the anthropology of science and technology, there is a growing literature on scientific environments (Forsyth, 2001; Gusterson, 1996; Helmreich, 1998, 2009; Knorr-Cetina, 1999; Latour & Woolgar, 1986; Rabinow, 1996, 1999; Traweek, 1988). The field of science and technology has a longer history in sociology (Bloor, 1976; Bourdieu, 1979, 2004; Collins, 1993; Lynch, 1985; Merton, 1973). Most of these studies, however, deal with science, ethics, gender, and other social issues but not with the workload or pay of scientists or with their working conditions more broadly. In sociology, there is also a field of literature that deals with labor issues (Abbott, 1988; Braverman, 1974; Friedson, 2001). The field of sociology of labor deals with labor in general and not with scientists specifically, and the core literatures do not deal with people's subjective perception of labor; it is more general in scope. This chapter, however, will try to complement this literature by emphasizing the subjective side, that is, life experience, of being a scientific laborer.

In the field of history, Shapin (1989) has written about the invisible technician at the time of the scientific revolution. All great names of the scientific revolution seemingly had workers helping them with experiments. These assistants were engaged to stay in a laboratory for a stated time, mostly a year at a time, to supply their labor at their master's command. There was a master–servant relationship, in which the servant was invisible and lacked identity in scientific matters. This has been especially true for women in science (Gould, 1997; Pyenson & Sheets-Pyenson, 1999). Today, there are still technicians in laboratories providing essential labor to the scientific enterprise. Nanoscientists often use cleanroom laboratories, in which microscopic particles are taken away in order to prevent interference with the atomic level experiments. These laboratories have a staff of technicians helping scientists and doing upkeep of the lab. Such technical staffs are often people with PhDs who are tired of research or graduate students who did not finish their degree.

In the social science literature, there is a cluster of literature that deals with high-tech workers in industry, a labor field closely related to the university scientists in this study. McKay (2006) follows high-tech workers in the Philippines. The workers are tied to 12-hour shifts in assembly lines to produce computer chips in harsh environments with anti-union bosses. Biradavolu (2008) in her study shows the importance of social networks among Indian "immigrant entrepreneurs." These entrepreneurs form a transnational capitalist class that influences work and working conditions in Silicon Valley. Slaughter, Campbell, Holleman, and Morgan (2002) investigate the collaboration between science and engineer faculties and industry in the United States. The authors argue that graduate students are "trafficked" as tokens of exchange between academia and industry. The argument is that leaders exchange subordinates to strengthen social network ties with other leaders, in a beneficiary win–win deal. Academia gives students to industry in return for equipment and funding. Invernizzi (2011) deals directly with nanotechnology and labor, and looks at the jobs and skill mostly needed in the private nanotech companies. The author argues that working with nanoscience will require a new set of interdisciplinary skills, a set of skills available only to the highly educated.

Unfortunately, there seems to be very little written about labor among university-bound scientists. Nanoscientists are highly educated, come from middle-class upbringing and work at universities. Despite that, it does not mean that everything is fine labor-wise. For example, a PhD student in nanoscience in Santa Barbara described her life in the following manner:

> They pay PhD students $1,500–2,000 per month. I'm also a teaching assistant, if you teach you do not have to pay the fee. Otherwise the fee is $3,000 per quarter, more than I get in salary. My rent is $1,000 per month and then utilities. PhD students here finish in five to six years. At work, I follow a strict schedule and need to plan work ahead. That is the only way to make the experiment to work, to do the right things on time. I have a close friend that is also a PhD student. We complain a lot together [laughter]. We see each other outside of work, usually once every second week. I do not see other colleagues after work. After work I go home, plan work, have a beer and watch TV. Weekends I hike by myself, I like to be outdoors. Not really time to do much . . . I do not have many options for a social life. Hard to find time with someone out of science.

She later told me that she estimated her average workweek to be about 60 hours and that she often had to come to the lab on Saturdays, too, to get research done, and this research was suffering due to the teaching load. A European postdoc described his life in Santa Barbara in the following manner:

> Here in Santa Barbara I don't have much time left. I have no extra job in Santa Barbara, but I tutor people [back home]. I have not been to one party here in Santa Barbara. I work a lot and I have no social life here. A lunch with colleagues occurs rarely. I have not made an effort here either of being social. I just don't have time.

The estimated workload of the postdoc was hard to predict. He told me that he preferred to work at nights, so he could be in touch with fellow colleagues back home, to whom he felt professionally close. A third example of the hard work was one evening when a scientific friend of mine was to pick me up. I was in Old Town Goleta, the town closest to campus. My friend lived in Goleta but had no idea that the central part of Goleta was named "Old Town," after four years living in the town. When I asked about it, I was told she mostly traveled to work and home, not really stopping or paying attention to what was in between.

My point here is that the 60–90 hours of workweek experienced by many of the nanoscientific workers whom I have studied in the United States and in Sweden limit their social and spatial life. Even if the researchers do come from all over the world, they spend so much time in their labs that they do not really have time to interact with the surrounding society. Instead, they limit their social interaction to colleagues, which enhances the importance of university-related social networks (Johansson, 2009).

There are also huge discrepancies in working conditions between the same types of researchers, depending on the source and relative abundance of research money. One method for reconciling these differences is to let one project that is well funded provide salary for another project that is less well funded, a method I believe is illegal both in Sweden and the United States. To my great surprise, when conducting interviews, I sometimes encountered people who did not work on nano at all, despite their official job descriptions. One PhD student told me bluntly, "[My professor] put me as an employee here to pay my salary. Don't really do things with nano." At first sight, this might seem like a rather ingenious method to acquire money for underfunded researchers. The PhD student, however, cannot make any official complaints about the work situation, as he or she officially is working on something else. There are also researchers who are funded by their home country and then do research in Sweden or the United States. Officially, they are paid but sometimes, due to corruption or other circumstances, the money stops coming. In Sweden, I met one researcher from Africa who was to get funding from home, but his money stopped coming due to corruption. His situation became really bad, and he had to get a second job, apart from his full-time research. After a while, his Swedish professor arranged to pay his rent.

Probably the worst form of pay for a PhD student or postdoc is a stipend. A stipend means that someone, usually the home university, pays money for a student to study somewhere else. In theory this sounds really nice, but in reality this can become a way for a research group to acquire workers for free and without obligations to provide adequately for them. In the United States, where one has to pay for healthcare, I have encountered researchers with stipends who are without healthcare coverage.

The common method of pay, however, is that the employer follows the rule. There are national differences between Chalmers University of Technology in Sweden and the University of California at Santa Barbara when it comes to salaries. Making international comparisons of salaries is tricky as the value of salaries for the individual should be contextualized to relevant cost of living. In Sweden, taxes are higher and eating-out and clothing is more costly than in the United States, whereas rent, healthcare, and telecommunication is cheaper. In Sweden, PhD students are paid employees for a four-year period, and in 2009, they earned approximately $41,000 per year. Postdocs earn a couple thousand dollars more. In Santa Barbara, in contrast, there is more variety in salaries as they are dependent on the source of funding. In my fieldwork, people told me PhD students in nanoscience, earned roughly $20,000–$24,000 per year in 2009. A U.S. postdoc in the chemistry department in 2009 earned $35,500 per year. If one is to consider work weeks of 60 hours, the hourly pay, both in the United States and Sweden, is not really that great.

Graduate school in science and engineering fields takes about five to six years at both universities, and a postdoc position lasts about two to four years. This means that the average researcher spends approximately seven to ten years of their working life with a low pay and a 60-hour or more workweek. I have also interviewed people who are into their third postdoctoral appointment, which means that the postdoc period can be extended to 10 years or more, making this "transitory postdoc period" more or less a permanent position inside the academy. Professors are rarely in the labs, so technicians, graduates, and postdocs do the majority of the hands-on nanoscientific work at universities. It is here that a conundrum exists: Nanoscience is a well-funded field of science, yet the average nanoscientific worker is not particularly well paid and works long hours.

Working long hours and being surrounded with toxic substances also creates working environment hazards (see Conti et al., 2008). After spending years doing fieldwork with scientific researchers, I have been surprised by how relaxed they are around these highly toxic materials. Accidents do occur and are rarely reported unless there are major accidents. The general attitude is that risk is part of the job, and if you are afraid, you are not cut out to work with nanomaterials, thus enforcing the notion of not to complain about the working environment.

Figure 6.2 Hazardous materials are often stored in ordinary refrigerators. It is important that brought-along food is kept separate from toxic substances.

SOCIAL NETWORKS

When doing ethnographic fieldwork, there are, of course, some people whom you get to know better than others. One of the people I got to know better was an American PhD student with a family. At one time, he told me that his professor wanted him to go to a conference, and he was not that keen on attending. After the conference, I asked him about the conference, and he told me:

> The main reason to go to conferences is to learn new things they tell you, this is not true. My interpretation is that it [conferences] is more of a social thing. There is yet not much time to get to know people. During dinner I sat with two people, but none of them did similar things as me. And then there is not much time for poster presentations . . . I like science, but not the social aspects of it. I have little time and the social time I got, I want to spend with my family. I do not see the social as part of my work, but I know it is. It is important when applying for job and stuff.

He confirmed what so many researchers have told me over the years. Being a nanoscientist is about science, not about being social. Yet, at the same time, social networks are important for furthering one's career (Bourdieu, 2004; Merton, 1973; Traweek, 1988).

When I lecture about social networks and scientists, I use the simile of an oasis in the desert. Each cleanroom lab is like an oasis surrounded with infertile dead land, from the scientists' perspective. Nanoscience can only thrive in these lab environments, and scientists move from one lab to another. These labs, even if placed all over the world, do form a rather exclusive and secluded social space. This spatial arrangement has been described as maintained by *intra-space mobility* (Mahroum, 2000). This means the scientists move globally between a limited number of research facilities. The reason to move is work related, as the individual scientists seek opportunities not available in the previous location. Usually there are, at the new location, better work/job opportunities, but it can also be as simple as the new location offering employment that did not exist at the previous location. Moving around also helps the individual researchers to expand their social networks with new colleagues, making connections with people whom they may later collaborate with. A professor in Sweden told me several years ago,

> To succeed, one needs both to be a good scientist and to build a broad network. The network is important when seeking funding. Everybody knows everybody in the business, and it is important to be considered serious to get funding. One must make a name for oneself inside the network.

The social networks of the scientists working with nano are not special, compared to other similar groups. Bourdieu (1979) introduces the term symbolic capital to understand how resources are distributed dependent on individual prestige. Prestige in the nanoscientific community is reached by publishing. The more recognized and publicized a scientist is, the easier it is to acquire financial means to continue research, which leads to more publications and further enhanced prestige—a process named the "Matthew effect" (Merton, 1973). In nanoscience, however, this individual prestige is reached by collaboration with fellow scientists, as research and publications are a collective endeavor, thus emphasizing the importance of maintaining social networks to reach success.

Beck and Beck-Gernsheim (2002) indicate three important characteristics of a contemporary labor market: *education*, *mobility*, and *competition*. Education leads, at least in theory, to better jobs than those jobs available without education. Among nanoscientists, the labor market often demands a change of residence, a change of social patterns, and a change of work. Competition among the scientists leads to an individualization and refinement of one's labor skills. On the nanoscientific labor market, the individual researcher sells his or her skills on a global market, trying to get ahead by not only acquiring better working skills but also by acquiring more extended social networks. One of the researchers told me, for example, how he got involved with a company he now worked for part-time:

> I got this job through personal contacts. I helped a professor to grade and teach students. We got to know each other. Later the professor asked me to join the company. This was during my graduate years.

Personal time and professional time often seem to blur when it comes to reaffirming and strengthening social networks among scientists working with nano:

> I know this guy who previously did a postdoc in a group here. He is now my boss at Intel. The two of us were on a bike ride when I was offered to work for Intel. The two of us are not friends but sometimes meet up to go dirt biking. My Intel boss wanted someone who knows the cleanroom.

When social networks, symbolic capital, and work skills are all meshed together, it is easy to see that complaining about one's labor conditions might hurt one's career. It is common for the employer to interview not only the worker but also colleagues, and to receive letters of recommendation. Frequently professors recommend prospective students to each other. When a person leaves a project, that person can recommend a peer to continue their work. In one of the groups, I followed a postdoc, who received a tenure-track job in academia and had to leave his position rather hastily. The postdoc recommended a colleague and friend of his from his previous university to take his vacated position. The advantage of this solution is that the project continued uninterrupted, as the new researchers were familiar with the predecessor's work. As they also were friends, it was easy for the new researcher to ask experimental advice from the prior researcher. As a consequence, the two researchers strengthened their professional social ties to advance their symbolic capital within the closely-knit community of nanoscientists.

My career as a social scientist interviewing nanoscientists goes a long way back. I was already interviewing scientists in 1999 as a master's student in Sweden. The purpose of my MA study was to examine if and how physics students obtained a new outlook on nature as a result of their field of study. One of the people I interviewed was a young student with a passionate interest in physics. Little did I know that we would follow each other throughout our careers. The physics student later enrolled in graduate school in nanoscience and completed her PhD about the same time as I finished mine. Looking back I find her insights, in a 1999 interview, regarding social networks in science, to be rather profound:

> The social networks are very important; I work better in a group. You get the answers together. Research is built on networks, which is the reason for conferences. It is important to fit inside the group and thus personal connections are important when applying for jobs. To get the highest degree on one's exams does not mean one is good at research. A feeling of connectedness is developed inside the group in which one works. This, though, does not suggest that one is personal friends [with those inside the group].

Working in groups to enhance one's personal career means that social connections are important for academic advancement. In a way, a scientific enigma is created. One's colleagues are one's competitors, as they are the ones with whom one is competing for jobs, lead authorship, and recognition from the group leader. However, one's colleagues are the ones who will enhance your career, as they are the ones that will assist you in your research, recommend you future employers, and help you with funding.

Universities follow a strict hierarchy, more so in the United States than in Sweden, but when it comes to advancing one's scientific career, the professors are gatekeepers, no matter the cultural setting. Jobs are often transferred between professors who know each other, and they do not want to place troublemakers among their peers, as this may make them look bad. I know of scientists who have not received good recommendations from their professors. For them, their career in academia is basically over, often necessitating moving on to industry instead. A positive recommendation, however, can be the gateway to a continued and successful scientific career. A PhD student in Sweden told me the following story in 2003:

> I did get a PhD position at Lund [University] but it later proved to have no funding. The professor in Lund, whom I should have worked for, thought that I should apply somewhere else. I searched the Internet and sent letters to interesting groups, all over Sweden. I got a reply from one of the groups at the Gothenburg [University] and was asked to send in my recommendations. I got a letter from the professor in Lund and I got the job.

This is a norm for job seeking/obtaining in the nanoscientific community. Recommendations from one's superiors are vital to a continued career inside the community, and thus most of the scientist I talked to do not want to anger their leaders by complaining about heavy workloads and long hours in the lab.

There are, of course, a few stories of resistance, where students and postdocs do revolt and through skillful social maneuvering, are able to switch supervisors. In general, this means that they also switch social networks. One of few PhD students I have interviewed who has successfully switched supervisors told me,

> It is important to put people as authors. I was cheated upon as an undergraduate. In a paper with my former supervisor, I had done two out of five diagrams and was not named an author. I got mad and told my supervisor, who got mad and yelled at me. I complained to the dean, and after some more argument, I was finally placed under the supervisor I have today. The reason why I work on the stuff I work on today is because of my current supervisor.

How this revolt will influence the future career of the scientist is hard to tell as the career is in the making. The PhD student in question wanted after graduation to quit science and go into teaching, a career choice that may have influenced the choice to pick a fight instead of being quiet. The general notion

Figure 6.3 Creating a private space in the lab. Squeezed in between lab equipment are some private pictures of someone's cats. Working long hours creates a need of feeling at home at work. On the right side of the wall, there is a sticker "I'm a grad student. Appreciate Me!"

inside the community, however, is not to pick a fight with one's supervisors, as it often ends badly for the researcher.

A CULTURE OF ENDURANCE

During my fieldwork in Sweden in 2003–2004, I followed an undergraduate class in nanoscience for physics students. At the end of a class which dealt with biomimicry in nanoscience, the teacher turned to the students and said,

> You will belong to the most 5 percent well-educated in Sweden. There will be an ethical discussion surrounding it [biomimicry in nanoscience]. To be able to have an opinion, you have to know what it is. Be confident that you have knowledge; then, you have an obligation to teach that knowledge to the general population.

At the end of the class, the teacher concluded that it is collegial competition that pushes science forward. One gives secondary knowledge to others and keeps the top knowledge for oneself for patent and articles. "It is tough" he said, "I do not recommend a scientific career unless one is fond of teaching and has something one really wants to discover." This understanding that science was tough and that, through their education, the students had joined an elite group was reinforced time after time, not only in class but also in the lab.

Some of the people I interviewed, both in Sweden and in the United States, even choose a career in nanoscience *because* they see it as a challenge. In response to the standard question as to why they chose a career in science, I often got comments such as "I chose environmental engineering as I loved math and science. [I] wanted a challenge and to study in a challenging environment; engineering was that" and "I needed to decide a major, wanted a challenge, and decided physics."

Science is not only mentally challenging; it is also challenging as it monopolizes the student's time. Spending long hours in school and in labs does seem to create what I would call "a culture of endurance"—a culture with lots of weeding out and dropouts. In Sweden, I was told that out of the 60 students enrolled in the physics program, about 15–20 students were still enrolled in the final year, and not all of those would graduate, which makes for a very high rate of attrition. This weeding-out process with many dropouts does reinforce an aura among researchers that they belong to a special group of people that has endured and is enduring hardships. This, in turn, helps to create a close-knit community as has been investigated by Hasse (2002) in her fieldwork among physics students in Denmark.

At one time, I followed a tour of professors, postdocs, and graduate students from Los Angeles. They were in Santa Barbara to look at some of the labs in which I had done my fieldwork. In one of the labs, the researchers showed us how they used microscopes for counting cells. Cell counting is a tedious process of studying cell deaths in a sample over a period of several hours. Sometimes it means that the scientists need to be awake for 24 hours continuously conducting a cell count every hour. In this case, they added nanoparticles into the cell compound to see if the cells were affected. We all took turns looking at the cells through the microscope when suddenly one of the professors from Los Angeles told the students that he had done cell counting as a graduate student and that cell counting was what separates real bio scientists from the rest. If one were able to put up with the tedious work of cell counting, one was fit to have a career in science, according to the professor. Similar stories of hardship, including working long hours and being awake for a long period of time, are commonly related by professors and senior researchers to their junior colleagues, thus transferring, fostering, and reinforcing a culture of hardship, a process that also adds to the exclusiveness of belonging to the closely knit social network of nonscientists.

Many of the experiments are conducted at odd hours as some machines are in huge demand, and some experiments run for several days, requiring constant attention. During my fieldwork, I collected data about what constitutes a typical workday, and scientists in both Sweden and the United States told me similar stories:

> During the day, there is no real break. No real lunch, I eat apples and carrots during the day. But no real lunch, lunch happens when there is time. I do not plan lunch nor go to working lunches. I use the waiting time between steps in the experiment to go to the bathroom, but there is rarely more time than 5–10 minutes. Sometimes there is a 40-minute break, and then, I sometimes go for a walk. I always prepare for the next step in the experiment.

Experiments that need constant overseeing do put limitations on time for lunches, bathroom visits, and time for social connections. The scientist quoted above also told me that social networking was mostly done through e-mails as those could be replied to when there was time between experimental procedures.

Workweeks of 60–90 hours are common among scientists dealing with nano, some without any vacation time. A female single parent, for example, told me that it is really hard to work 70 hours per week and to have a family; she also told me that was the reason why there are so few single female parents in science. It is difficult to combine parenting with the massive working hours. The long working hours do affect people's social life. There is a need to find a partner who can cope with the scientist's odd hours and all-consuming work schedule.

When I first arrived to Santa Barbara, I did not have many friends and participated in the reactivation of the postdoc organization at the university.

The vast majority of the postdocs on campus were engineers and natural scientists, and thus, I involuntary got a new channel of information as some of the postdocs working with nano participated in the postdoc organizations activities. At one time, a European female postdoc active in the nano field, when hearing about my work, said that I needed to write the truth about scientists. I was curious as to what "the truth about scientists" was; she said that the truth was that science was unequal for females. She continued by saying that as a male scientist you can work hard now and marry later. You can postpone child rearing until later in your career, when you are a professor. Then, you can always marry a young female student, she told me. She said, "Female scientists don't have that option. If female scientists want to have kids, it will affect their career." She also reflected on how often female professors active in her field were without kids. I contradicted this notion by saying that there just are not that many female professors in the field of nanoscience, period. The female postdoc had no child or spouse of her own, saying there was no time for it. She just worked too hard. Later, her professor had let her go with short notice. In two weeks, she had found a new postdoc position and left Santa Barbara to continue her migratory life as a nanoscientist.

Finding nanoscientists who do not continue their academic careers is difficult as they no longer are part of the university system. It is thus much easier to study those who have tolerated the culture of endurance than to study those who dropped out. After some investigation, I did encounter a French woman, who after completing her PhD, decided to stop being a scientist. She told me,

> They [scientists] are a weird bunch. Some of them are really antisocial and only think of research. To work with research also means that you work all the time, there are no 40-hour workweeks. There is also too much competition for funding. To continue research would have meant I would become a lab rat, spending all my time in the lab. It felt like I missed out on the rest of life.

Stories of hardship and endurance and the weeding out of those who do not put up with it create a culture in which one does not complain. Those who complain just are not cut out to be scientists. There is also a general notion that to become a good scientist, one needs to work a lot and sacrifice one's social life. One scientist said bluntly, "I have no need to be the best scientist. I also like [to] have a life!" The community of nanoscientists, like all communities, thus consists of people that adapt to the cultural norms. The French woman, mentioned already, did not want to adapt and left the community. This is a weeding out that starts early. Hasse (2002), in her fieldwork among university physics students, shows a strong weeding-out process among the students; some adapt and thrive and others do not and are weeded out. As there is plenty of group work, the students have to fit in to find lab buddies. Kvande (1999), in her work among female engineers, describes different strategies that females take to adjust to a male-dominated engineering workplace. One of the strategies is sameness in which female employees do not see themselves as any different from their male counterparts. During my fieldwork in Sweden (Johansson, 2009), I found the sameness ideal to be the norm. The few females who were active in nanoscience told me they had worked with men their whole career and thus, were used to it. They did not see themselves as any different from the men, and those who thought they were different were weeded out.

The culture of endurance also influences the scientist's risk-related behavior. Stories of getting burned by Bunsen burners or chemical compounds that explode are told and retold among one's colleagues, adding to the notion of nanoscience as a tough line of work. People have told me that lab workers live generally 10 years less than the average population, whereas others told me they live longer than average as they build resistance to chemicals. Both of these views represent folk models of risk that enhance the notion of a culture of endurance. Risk at work is often not directly attributed the source, that is, chemicals or machines, but more to the handling practices of people in the laboratory workplace. A postdoc working frequently in the cleanroom told me,

> There are risks with cleanroom work, psychologically. I work through the nights, sometimes for weeks. That can affect your outlook on life, and you can become depressed. You are around chemicals and machines; you need to be aware of your surroundings. Cryogenics is probably the largest threat. You don't think about risks; they are just part of your surroundings.

Once more, a culture of endurance is enforced in which those who are not able to endure the long working hours enhance risk. The enemy is not the dangerous things in the surroundings but the scientists themselves who need to be psychologically stable to cope with the long and tedious hours in the lab. Another scientist, when asked about potential risks of work, concluded, "As a scientist, you cannot work if you are afraid. This is our job."

THE "TRANSITORY PHASE OF HARD LABOR" MYTH

Nanoscience in academia is a rather strict meritocracy in which undergraduates become graduates, who get their master's degree, who then get their PhDs. Then they continue as postdocs, who become assistant professors, who finally become full professors. In each step on the way to full professorship people drop out, as there are more people in the previous stage than there are in the following stage. Yet, it is interesting to note that there is a general idea that all the previous stages, before full professorship, are seen as transitory positions or steps. The majority of the scientists I have talked to, both in the United States and in Sweden, over the years are on their way to do something else, either getting a new position or moving to a new university. It is interesting to note, however, that when asked about the future, the scientists often respond that they want to continue to do research or teach; only rarely do people say that they aim to become full professors. A certain type of humbleness and taking one career step at a time seems to be the norm, probably as a self-protection against not reaching the goal.

It is thus commonly acknowledged that the hard work and low pay are just temporary and that better times are to come. At one time in 2009 I was having a conversation with a fellow postdoc regarding the low salaries in chemistry. A union for the postdocs had just been formed at the University of California system, and one of the main issues they pushed was higher minimum wages. To my great surprise, my postdoc colleague told me that he was against the unionization of postdocs. His reason for it was that if salaries rose and the postdocs got better working conditions, then people would stay as postdocs; they would not move on in life.

The problem is, of course, that many scientists already get "trapped" in a transitory phase between having a PhD and becoming full professors in nanoscience, as the number of full professorships is limited. In each research group, there is only one full professor at the top. Being a postdoc and researcher for 10 years or longer is common among the nanoscientists, and for several, it becomes a permanent position in their academic career. Both in Sweden and in the United States, there are many foreign-born researchers who do not want to return to their home countries, thus constituting a more or less permanent pool of "transitory phase" laborers. Taking another postdoc becomes a way to stay in the United States or in Sweden even if the working hours and pay are not satisfactory. An Asian postdoc, for example, described his situation:

> This is my third postdoc at UCLA. I'm from [home country], but my I got my PhD in USA. USA is a good place for kids. I don't want to go back, and I'm looking for an academic position in USA, but it is hard . . . I have to choose my career after the jobs that are available, so I ended up here by chance. It is hard to survive as a scientist. I like the academy. I want job security and the academy is good even if salary is less [compared to private industry].

Other foreign researchers only see their work as transitory and have a wish to return home:

> Given the choice, I would like to go back to [home country] and get an assistant professorship. They have a culture I can identify myself with, better than USA. [home country] is better, it's easier to get grants [there] and easier to run a lab. It is also easier with visa and green

> cards. [home country] is Asian culture, I know how to deal with students there and I know how to handle things. Maybe it will get better here with time. My wife is from [Asia], she is working here in the USA, working in a lab. I have a contract for three years. We'll see what happens after that—it depends on the opportunities.

Wanting to stay in the United States means that foreign researchers accept long working hours and often low pay, whereas foreign researchers who plan to leave may also accept long hours and low pay, because they see their position as only temporary. Complaining about salary or long working hours may brand the worker as a troublemaker, which thus makes it hard for them to find a new position, as basically all positions except full professorships are temporary.

The so-called transitory phase of hard work also takes its toll on private life, because moving around and working long hours may affect partners or spouses adversely. A scientist in Sweden who was just finishing his PhD told me about his future prospects:

> Many PhDs don't know what they want. Some of them are unemployed but several seem to have some kind of plan. One girl I know now studies to become a minister. Another friend of mine has a hard time. He got a prestigious postdoc in USA at a top university. It was hard on his relationship with his wife though, and they divorced. As he did not want to lose custody of his kids, he had to terminate his postdoc in USA and return to Sweden. Now he works part-time at our department. In physics, it is standard to do two postdocs in two different countries. After that, people seem to return to Chalmers [University of Technology].

The transitory nature of nanoscientific work does not only monopolize scientists' time. It also monopolizes their spatial mobility, as it is a norm to move between universities and countries in different phases of the career, moves that may cause personal sacrifices for the scientists and the people close to them. It is only in the professorial stage of labor that a more permanent place to live is achieved, thus making the transitory phase a period of great uncertainty socially, monetarily, and geospatially.

CONCLUSION

In popular media everywhere, it is usually the scientific elite that is seen and heard. It is Noble laureates and professors from top universities who enlighten the public on scientific matters. They are usually also well funded, with a staff of scientists supporting them in the scientific work on the bench. This is far from the reality of average scientific work conducted at the majority of the world's universities. In a way, the scientific elite can be likened to the elite in professional sports such as NBA basketball. A few basketball players earn millions of dollars from their sports, whereas the majority of the players in different leagues are struggling financially while living anonymous lives. In the media, however, the basketball elite is talked about and referred to and thus form the norm of what it means to be a basketball player in the public's eyes.

The scientific elite are not only more often seen in public media, they also have the capacity to increase their economic and social capital (Bourdieu, 2004). This cumulative advantage in science is often referred to as the "Matthew effect." It is inspired by a passage from the Gospel of St. Matthew: "For unto every one that hath shall be given, and he shall have in abundance: but from him that hath not shall be taken away even which he hath" (Merton, 1973, p. 445). In the realm of nanoscience, it is the professors who are the masters of their research group. It is they who employ and fire people; it is they who decide which field of science they want their group to pursue; and it is they who often receive the scientific, economic, and social rewards for the group's work.

In this world of professor-controlled research groups, it is easy to see how the present scientific culture of labor among nanoscientists has formed. Professors are important gatekeepers for the careers of upcoming scientists, and thus, the younger generation has nothing to gain by complaint or to risk being seen as a troublemaker. The professors and other senior researchers themselves work long hours. They are the people who persevered through their own hardship of long lab hours when they were younger, and they tell these stories to the younger generation. Thus a *culture of endurance* is formed, and those who cannot endure or want another lifestyle leave. Among the young generation of rising scientists, there is an understanding that, if successful, they will one day join the ranks of the senior researchers, to become professors and maybe even join the scientific elite. This understanding of social progression in the nanoscientific community is a powerful motivation for enduring hardship that is frequently perceived as just temporary. The dream of one day joining the scientific elite or becoming a professor is for most scientists just a dream, as each research group is led by only one professor.

To understand the connection between social network dependencies, a culture of endurance, and a perception of a linear career and to understand why scientists do not complain about their work environment, one must examine the cultural economy among nanoscientists. Each community has its own capital. What is considered prestigious in one community may not be considered prestigious in another, a phenomenon Bourdieu (2004) named "symbolic capital." For example, what is considered prestigious and gives high rank among criminals is often looked down upon by civil society (Bourgois, 2003). Prestige and high rank in the nanoscientific community is reached by publishing scientific results. In their influential laboratory study, Latour and Woolgar (1986) use economic metaphors to understand the scientists' use of publications as a base for prestige. Scientific publications are accumulated, shared (by adding authors), and can even be stolen (by publishing other people's results), just as money can in civil society.

In nanoscience, researchers publish with peers and are thus dependent on other scientists to accumulate articles. One way to anger collaborators is by not doing one's share, such as working fewer hours than others and complaining about the workload, complaints that do not lead to publications of more articles and higher prestige. The cultural economy of nanoscience and the dependence on social networks are closely intertwined.

The cultural economy of nanoscience fosters quite naturally a culture of endurance. Those people who succeed in the culture, by becoming successful professors and role models, are those who work long hours and make personal sacrifices for science. These are the same people who work as gatekeepers, selecting those "who have it" from those "who don't" when it comes to scientific careers, thus transferring the code of community work ethics to the next generation of scientists.

The perception of linear careers is also connected to the cultural economy of nanoscience in which articles are accumulated, and with time, the scientist, if he or she accumulates enough merit, will join the ranks of the community elite. The understanding that success comes through hardship and sacrifice is a powerful universal myth. Campell (2008), in his work on comparative mythology, follows the archetypical hero. All over the world, mythological heroes seem to follow the same structure. They start out in the ordinary world, start a quest, and enter a new strange world, endure hardship, return to the ordinary world with gifts or new talents. Nanoscientists make a similar journey. They start out in the ordinary world as laypersons, start a scholarly quest to join the community in which they study the strange world at nanometer scale. During years of study, they endure hardship, such as working long hours with low pay. They endure all this so they can later return to the ordinary world as successful scientists and distinguished members of the nanoscientific community. The myth serves a great purpose to make people work hard inside the community, to give hope of a better future, and to cement the culture of endurance. But in reality only a few reach full professorship.

In a sense, nanoscientists are not particularly different from other late industrial middle-class workforces (see Beck & Beck-Gernsheim, 2002). Middle-class occupations are more socially mobile with both upward and downward mobility, dependent on individual performance (see Lawrence, 2006; Sulkunen, 1992). The nanoscientists live in a meritocracy in which one's personal skills are sold on a free market. The market in this case is different research groups. The personal skills are used to increase one's merit inside the community, and with high merit, comes access to more prestigious research groups. As nanoscientists work in groups, also one's colleagues' work performance is important for the accumulation of individual merit. People who don't accept the long work hours not only destroy their own careers; they may also hurt their colleagues' careers. Each individual must accordingly be aware of the performance of those he or she works with and avoid those who do not live up to the community standard. Those who do not work late, and those who complain, will have a hard time finding collaborators, and in the end, they will be forced out of the community as they do not comply with the community norms of endurance and no complaints.

There is also a more general "scientific ethos" that emphasizes self-invisibility. According to Haraway (1997), the influential writer of science, the self-invisibility found among scientists is a modern, European, masculine form of modesty in which the scientist is believed to add nothing to their observations. This self-invisibility becomes the foundation of objectivity in which the scientist just bears witness to nature. The ethos of self-invisibility may also be understood in that the scientists are not to complain about their workload or working conditions, as this would increase their subjectivity and add personal points of view to the scientific process.

So, even in the hyped-up and well-funded research field of nanoscience, there are highly educated, highly skilled technology workers, who do a lot of difficult and laborious work without receiving recognition or rewards. For them, working *with* nature's next to nothing, the realm of atoms and molecules, also means that they are working *for* next to nothing.

ACKNOWLEDGMENTS

I am in gratitude to the editors of this volume, Barbara Herr Harthorn and John Mohr, for their valuable comments. I would also like to thank my colleagues at the Center for Nanotechnology in Society at University of California at Santa Barbara for assisting me in the process of writing, especially Cassandra Engeman, with whom I discussed labor issues. Ann-Sofie Sten helped me to format the pictures.

This material is based upon work supported by the National Science Foundation under Cooperative Agreement No. SES 0531184 to the Center for Nanotechnology in Society at University of California at Santa Barbara. Any opinions, findings, and conclusions or recommendations expressed in this material are those of the author and do not necessarily reflect the views of the National Science Foundation.

REFERENCES

Abbott, A. (1988). *The system of professions: An essay on the division of expert labor.* Chicago: University of Chicago Press.

Abu-Lughod, L. (1991). Writing against culture. In R. G. Fox (Ed.), *Recapturing anthropology: Working in the present*, pp. 137–154. Santa Fe, CA: School of American Research Press.

Beck, U., & Beck-Gernsheim, E. (2002). *Individualization: Institutionalized individualism and its social and political consequences.* London: Sage Publications.
Berube, D. M. (2006). *Nano-hype: The truth behind the nanotechnology buzz.* Amherst, MA: Prometheus Books.
Biradavolu, M. (2008). *Indian entrepreneurs in Silicon Valley: The making of a transnational techno-capitalistic class.* Amherst, MA: Cambria Press.
Bloor, D. (1976). *Knowledge and social imagery.* Chicago: University of Chicago Press.
Bourdieu, P. (1979). *Distinction: A social critique of the judgment of taste.* London: Routledge & Kegan Paul.
Bourdieu, P. (2004). *Science of science and reflexivity.* Cambridge: Polity Press.
Bourgois, P. (2003). *In search of respect: Selling crack in El Barrio* (2nd ed.). Cambridge: Cambridge University Press.
Braverman, H. (1974). *Labor and monopoly capital: The degradation of work in the twentieth century.* New York: Monthly Review Press.
Campell, J. (2008). *The hero with a thousand faces* (3rd ed.). Novato: New World Library.
Collins, H. (1993). *Artificial experts.* Cambridge, MA: MIT Press.
Conti, J. A., Kilipack, K., Gerritzen, G., Huang, L., Mircheva, M., Delmas, M., Harthorn, B. H., Appelbaum, R. P., Holden, P. A. (2008). Health and safety practices in the nanomaterials workplace: Results from an international survey. *Environmental Science & Technology, 42*(9), 3155–3162.
Forsyth, D. E. (2001). *Studying those who study us: An anthropologist in the world of artificial intelligence.* Stanford, CA: Stanford University Press.
Friedson, E. (2001). *Professionalism: The third logic.* Cambridge: Polity Press.
Frykman, J. (1986). Hur fostras en svensk intellektuell? *Tvärsnitt, 8,* 54–61.
Giddens, A. (2001). *Sociology.* Cambridge: Polity Press.
Gould, P. (1997). Women and the culture of university physics in late nineteenth-century Cambridge. *British Journal for the History of Science, 30*(2), 127–149.
Gusterson, H. (1996). *Nuclear rites: A weapon laboratory at the end of the Cold War.* Berkeley: University of California Press.
Haraway, D. J. (1997). *Modest_Witness@Second_Millenium.FemaleMan©_Meets_ OncoMouse™.* New York: Routledge.
Hasse, C. (2002). Learning physical space: The social designation of institutional culture. *FOLK: Journal of the Danish Ethnographic Society, 44,* 171–194.
Helmreich, S. (1998). *Silicon second nature: Culturing artificial life in a digital world.* Berkley: University of California.
Helmreich, S. (2009). *Alien ocean: Anthropological voyages in microbial seas.* Berkley: University of California Press.
Invernizzi, N. (2011). Nanotechnology between the lab and the shop floor: What are the effects on labor? *Journal of Nanoparticle Research, 13*(6), 2249–2268.
Johansson, M. (2009). *Next to nothing: A study of nanoscientists and their cosmology at a Swedish research laboratory.* Gothenburg: Gothenburg University.
Keesing, R. M. (1994). Theories of culture revisited. In R. Borofsky (Ed.), *Assessing cultural anthropology*, pp. 301–310. New York: McGraw-Hill Inc.
Knorr-Cetina, K. (1999). *Epistemic culture: How the sciences make knowledge.* Cambridge, MA: Harvard University Press.
Kvande, E. (1999). "In the belly of the beast": Constructing femininities in engineering organizations. *European Journal of Women's Studies, 6*(3), 305–328.
Latour, B., & Woolgar, S. (1986). *Laboratory life: The construction of scientific facts* (2nd ed.). Princeton, NJ: Princeton University Press.
Lawrence, J. (2006). *The middle class: A history.* London: Little, Brown.

Löfgren, O. (1987). Deconstructing Swedishness? Culture and class in modern Sweden. In A. Jackson (Ed.), *Anthropology at home*, pp. 74–93. London: Tavistock.
Lok, C. (2010). Small wonders. *Nature, 467,* 18–21.
Lynch, M. (1985). *Art and artifact in laboratory science.* London: Routledge and Kegan Paul.
Mahroum, S. (2000). Scientists and global space. *Technology in Society* 22(4), 513–523.
McKay, S. C. (2006). *Satanic mills or silicon islands? The politics of high-tech production in the Philippines.* Ithaca, NY: Cornell University Press.
Merton, R. K. (1973). *The sociology of science: Theoretical and empirical investigations.* Chicago: The University of Chicago Press.
Pyenson, L., & Sheets-Pyenson, S. (1999). *Servants of nature: A history of scientific institutions, enterprises, and sensibilities.* New York: W. W. Norton & Company.
Rabinow, P. (1996). *Making PCR: A story of biotechnology.* Chicago: University of Chicago Press.
Rabinow, P. (1999). *French DNA: Trouble in purgatory.* Chicago: University of Chicago Press.
Shapin, S. (1989). The invisible technician. *American Scientist, 77,* 554–563.
Slaughter, S., Campbell, T., Holleman, M., & Morgan, E. (2002). The "traffic" in graduate students: Graduate students as tokens of exchange between academia and industry. *Science, Technology & Human Values,* 27(2), 282–312.
Sulkunen, P. (1992). *The European new middle class: Individuality and tribalism in mass society.* Aldershot, UK: Avebury.
Traweek, S. (1988). *Beamtimes and lifetimes: The world of high energy physicists.* Cambridge, MA: Harvard University Press.

7 Nanotechnology as Industrial Policy
China and the United States

Richard P. Appelbaum, Cong Cao, Rachel Parker, and Yasuyuki Motoyama

Since 2000, when the United States officially launched its National Nanotechnology Initiative (NNI), global public spending on nanotechnology has totaled an estimated $67.5 billion. If one includes corporate research and private funding more generally, the total of public and private spending is predicted to reach as much as a quarter of a trillion dollars by 2015 (Cientifica, 2011). Clearly, pubic officials across the world have come to see nanotechnology as the next technological revolution; firms and investors—no doubt in part attracted by the availability of public funding—have followed suit. Does this race to the bottom—investing significant public resources in nanoscale-level research development, and possibly commercialization—constitute industrial policy? How successful is it likely to be? A comparison of the two countries that have invested the most in nanotechnology will hopefully shed some light on these questions.

SCIENCE, TECHNOLOGY, AND INDUSTRIAL POLICY

In his classic work, *MITI and the Japanese Miracle: the Growth of Industrial Policy*, Chalmers Johnson (1982) made the now-classic distinction between "plan-rational," "market-rational," and "plan-ideological" state approaches to industrial policy:[1]

> In the plan-rational state, the government will give greatest precedence to industrial policy, that is, to a concern with the structure of domestic industry and with promoting the structure that enhances the nation's international competitiveness. The very existence of an industrial policy implies a strategic, or goal-oriented, approach to the economy. On the other hand, the market-rational usually will not even have an industrial policy (or, at any rate, will not recognize it as such). Instead, both its domestic and foreign economic policy, including its trade policy, will stress rules and reciprocal concession. (pp. 19–20)
>
> Economies of the Soviet type are not *plan-rational* but *plan-ideological*. In the Soviet Union and its dependencies and emulators, state

> ownership of the means of production, state planning, and bureaucratic goal-setting are not rational means to a developmental goal (even if they once may have been); they are fundamental values in themselves, not to be challenged by evidence of either inefficiency or ineffectiveness. (p. 18)

Johnson's tripartite distinction of policymaking was based on two interacting dimensions: the principal type of economic governance (market driven vs. state planning) and the principal type of decision making (ideologically driven vs. what might be today called "evidence based"). In addition to the crudeness of the resulting binary distinctions, Johnson's framework is missing a logical fourth category: "market-ideological." As Henderson and Appelbaum (1992) reformulate Johnson's original typology,

> [In] *market-ideological* political economies . . . public policy is oriented above all toward assuring free market operations. Like plan ideological political economies, market ideological regimes arise from ideological dogma: in the case of the former, the wisdom and benevolence of state managers in a command economy; in the case of the latter, the wisdom and benevolence of an invisible hand in a supposedly unfettered market.
>
> We refine the definition of *market-rational* political economies to include the regulatory function of the state, which is viewed in such political economies as providing a framework wherein investment, production, and distributional decisions (which remain the preserve of business) can operate in a relatively efficient manner. (p. 19)

Chang (1994) subsequently emphasized the state's engagement in "institutional adaptation and innovation to achieve goals of long-term growth and structural change" (p. 33), whereas Woo-Cumings (1999) incorporated similar notions in characterizing industrial policy as "the ability of the state sector both to accommodate itself to the changing requirements for remaining competitive in the global market place and to provide support for educational infrastructure and for research and development" (p. 27). O'Riain (2004) pointed out a facilitating role played by the states of Israel, Ireland, and Taiwan, such as fostering international networks and establishing venture capital funding and innovation centers. In the area of technology, industrial policy can take the form of what have been termed "horizontal technology policies" (HTPs)—policies that involve a class of subsidies that employ market mechanisms and self-selection to advance particular technologies (see, e.g., Breznitz, 2007; Hall & Rosenberg, 2010; Teubal, 1997). This aspect of industrial policy is similar to Block's (2011) analysis of the expansive role that governments can play in institutionalizing technological capacity.

The notion of industrial policy, in the plan-rational sense originally developed by Johnson (1982), is sufficiently broad to embrace a wide range of state-supported activities intended to advance economic growth and global competitiveness. Industrial policy could range from state policies intended to broadly support economic growth (e.g., providing infrastructure, support for science parks, subsidizing higher education, giving tax credits to research and development investment) to "picking winners" (typically targeted industrial sectors, specific industries, or even promising firms). In an effort to narrow the concept and adapt it to current conditions,[2] economist Dani Rodrik (2004) proposes that a "twenty-first century industrial policy" would involve "strategic collaboration between the private sector and government with the aim of uncovering where the most likely obstacles to restructuring lie and what types of interventions are most likely to remove them" (p. 38). In Rodrik's formulation, the government does not pick particular sectors; rather, it provides support for activities that seem likely to enhance economic advancement—for example, promising frontier technologies. For Block (2008b), this suggests that industrial policy should involve "four distinct but overlapping tasks—targeted resourcing, opening windows, brokering, and facilitation" (p. 172). Industrial policy is thus a form of "embedded autonomy" involving

> a delicately balanced combination of (1) capable, coherent bureaucracies characterized by meritocratic recruitment, long-term career rewards, and high espirit de corps, with (2) dense ties to industrializing elites, which provides access to information, agents for implementation, and catalyzes a more coherent, forward looking entrepreneurial class. (Peter Evans, as quoted in L. Evans, 2004)

The Rodrik-Block-Evans version of industrial policy makes two important assumptions: that to be effective, industrial policy must involve those government agencies that have proven competence and effectiveness (indeed, that such agencies even exist), and that the twenty-first-century turn toward flexible accumulation and organizational decentralization—what Castells (2004) terms the "network society"—favors the kinds of limited government identified by Block (2008b). Whitford and Shrank (2011) take this argument one step further. Drawing on institutional theories of governance that emphasize the centrality of "networks that are neither market nor hierarchy" (Powell, 1990, p. 301), they argue that an effective form of industrial policy would play the role of "network governance." Although acknowledging that "network failure" can be a significant problem (and one that is too often ignored in the network governance literature), they argue that "network governance can in some situations foster learning, investment, and joint problem-solving far better than can alternative modes of governance" such as markets or hierarchies (Whitford & Shrank, 2011, p. 268).

To return to our initial distinction, building on Johnson's original formulation as modified by Henderson and Appelbaum (1992, p. 19), where do the United States and China fall in terms of the four quadrants based on their principal types of economic governance and decision making? Or is it more fruitful to view their approaches through the lens of facilitation, as O'Riain (2004), Block (2008b), Rodrik (2004), and Whitford and Shrank (2011) in various ways argue—subsidizing market mechanisms, supporting scientific activities believed to enhance economic development, playing the role of honest broker, and more broadly engaged in network governance?

In the next two sections of this chapter we review the history and current approach to fostering nanotechnology in the United States and China, after which we draw some conclusions with regard to their differing approaches to industrial policy.

U.S. INDUSTRIAL POLICY AND THE NATIONAL NANOTECHNOLOGY INITIATIVE[3]

Ever since the United States rejected the Japanese model of industrial policy during the market-oriented "Reagan Revolution" of the 1980s, "leaders in the legislative and executive branches who were concerned about US competitiveness took initiatives that helped the US to develop its own highly decentralized form of industrial policy that takes advantage of US global leadership in scientific and engineering research" (Block, 2008a, p. xx). This "hidden" or "stealth" approach to a quasi-industrial policy is consistent with the decisions leading up to the passage of the NNI. This indirect approach has fueled debates as to whether, and to what extent, the United States has an industrial policy at all. Although spending on defense is sometimes mentioned as a form of industrial policy,[4] Johnson (1982), as mentioned already, was generally skeptical—although he thought an industrial policy might be emerging in the 1970s. Historian Otis Graham (1992) has convincingly argued that the debate on industrial policy effectively ended with the Carter administration (1976–1980), and presidential candidate Walter Mondale, in his 1984 unsuccessful race against Ronald Reagan, abandoned the terminology altogether.

Robert Reich, who had been a strong advocate of industrial policy prior to joining the Clinton administration as Secretary of Labor between 1993 and 1997, described his inability to effectively implement any of his ideas in his humorously titled autobiographical memoir of the period, *Locked in the Cabinet* (1997). Reich had long been an advocate of the need for an activist U.S. industrial policy, in the face of rising East Asian and European economic power. In Reich's view (1997),[5] private–public efforts such as Europe's Airbus called for an equally aggressive response on the part of the United States, which also needed to retrain its workers for participation in a high-technology global economy. In a Politico.com commentary on the Troubled Asset Relief Program (TARP), the Bush–Obama response to the financial 2008 financial collapse, Reich (2009) ironically noted, "America now has a full-blown industrial policy—a combination of lemon socialism and taxpayer-funded regulation." He went on to argue that industrial policy is required "to fill in where the market fails—providing basic research to help spur new technologies and industries, reducing the negative side effects of the market (such as carbon pollution), and easing the adjustment of workers and communities out of older industries that are shrinking toward new ones."

Most scholars would agree that the U.S. approach to industrial policy, at least until the last quarter century, had been market-rational: It assumed that market signals generally trump governmental decision making, calling for a hands-off attitude on the part of government bureaucrats and regulators. The basic function of government (with the important exception of defense spending) was to offer the minimally necessary regulatory framework, provide basic infrastructure, and set the rules for competition (Hall & Soskice, 2001; Pempel, 1999; Wade, 1990). During the past quarter century, however, the United States has moved increasingly from being a market-rational state to a market-ideological one (see Henderson & Appelbaum, 1992). Markets are supposed to select and develop leading-edge technologies, and determine appropriate levels of investment. As Neal Lane, former head of the National Science Foundation (NSF) and one of the central architects of the NNI, characterized the view that shaped the emergence of the NNI,

> The appropriate role of the federal government in anything having to do with private industry is a politically contentious matter in our free market system. There are people who believe that the most important thing government can do to assure that US companies are competitive in the global marketplace is to get out of the way, cut taxes, and reduce regulations. That is the view of the G.W. Bush administration. (Lane, 2008, p. 259)

As Lane elaborated in a subsequent interview,

> We've done battles for as long as we've had anything called science and technology policy. Although we did believe that government support of S&T [Science and Technology] (including nanotechnology) would help industry, I don't remember any conversations when people used the phrase "industrial policy." That's a four letter word. (N. Lane, personal communication, August 20, 2009)

The NNI, however, provides an interesting hybrid case for this framing: It did not derive from a groundswell of private sector initiative, nor was it the result of strategic decisions by government officials. Rather, it resulted largely from the vision and efforts of a small group of scientists and engineers at the National Science Foundation and the Clinton White

House in the late 1990s, who orchestrated widespread stakeholder buy-in, with Congress playing a more passive role limited to budgetary oversight. The NNI was not born of market-driven forces; ultimately, Washington selected nanotechnology as the next big technological breakthrough, initiated the policy, built private sector support, and invested in its development on a multibillion dollar scale.

The driving force behind the 1999 creation of the NNI, Mihail Roco,[6] has described the effort as a "bottom-up approach . . . an *inclusive process* where various stakeholders would be involved . . . a 'grand coalition' of academia, industry, government, states, local organizations, and the pubic" (2007, pp. 3.5–3.6). By his account, beginning in 1996, Roco organized a series of working groups involving experts from academia, industry, and government to lay the groundwork for making a convincing argument to the White House. Core proponents eventually came to include Evelyn Hu, a physicist specializing in nanoscale electronics at the University of California at Santa Barbara;[7] Richard Siegel, a materials scientist at Argonne National Laboratories;[8] and somewhat later, Neal Lane, a physicist who played key roles first as NSF Director, then subsequently as the Assistant to the President for Science and Technology and Director of the White House Office of Science and Technology Policy;[9] and Tom Kalil, then a trade specialist with a background in political science and economics, who was Deputy Assistant to the President for Technology and Economic Policy, and Deputy Director of the White House National Economic Council.[10]

Kalil, who had previously obtained funding for the Next Generation Internet and other IT-related initiatives, contacted Roco as early as 1997 (Roco, 2007), when—as he reported much later in an interview—he had "started looking around for the 'next new thing' after the Internet. That's when I stumbled across nanotechnology [and] discovered that there was a grassroots group of program managers that had been meeting on nanotechnology" (as cited in McCray, 2006, p. 20). Lane, who was finishing up his directorship at the NSF, was asked at a Congressional hearing if he could identify one area of science where major breakthroughs were likely to occur within the next 10 years. When he pointed to nanotechnology, his prediction "was met with blank stares. No one knew what it was" (N. Lane, personal communication, August 20, 2009). Roco, Kalil, and Lane then launched cross-cutting working groups to study the potential impact of nanotechnology, which they viewed as an opportunity with immense potential to start a new program (Lane & Kalil, 2005). The workshop reports, which eventually provided the key arguments in support of establishing the NNI, described nanotechnology as having the potential "for revolutionizing the ways in which materials are produced and products are created" (WTEC, 1999, p. xvii). They warned that U.S. industry, and thereby the U.S. economy, would suffer if sufficient government investment in nanotechnology were not forthcoming. As one report argued, "[T]here is an evident need to create the infrastructure for science, technology, and facilities, and human resources," because such investment "will be a strategic need for the next century" (WTEC, 1998, p. 1).

Roco made his pitch at a March 1999 Office of Science and Technology Policy (OSTP) meeting, and, despite competing proposals for other science and technology investment priorities, the NNI was eventually selected for funding. President Clinton announced the initiative at Caltech in January 2000, and the NNI was launched with an initial budget of $489 million for FY 2001 (Roco, 2007; see also McCray, 2005). Four rationales were used to justify the formation of the NNI: scientific necessity, grassroots ("bottom-up") support, economic competitiveness, and foreign threat. The challenge was now to build grassroots support for this effort, because neither Congress nor most of the industries that were predicted to benefit from NNI investment had any knowledge of nanotechnology. Roco's characterization of his efforts as "a bottom-up approach" notwithstanding, it is clear that the principal impetus and direction—from background reports to budget scheme—came from the top. Building stakeholder support was clearly crucial. The time was ripe: Growing concern about U.S. economic competitiveness contributed to the NNI receiving a largely sympathetic reception at Congressional hearings on the topic. The President's Committee of Advisors on Science and Technology (PCAST)[11] prepared a letter in November 1999 claiming that nanotechnology

> will have a profound impact on our economy and society in the early 21st century, perhaps comparable to that of information technology or of cellular, genetic, and molecular biology . . . Investments in nanotechnology have the potential to spawn the growth of future industrial productivity. (1999a)
>
> [T]he country that leads in discovery and implementation of nanotechnology will have great advantage in the economic and military scene for many decades to come. (1999b, p. 2)

Significantly, PCAST also argued that the private sector, by itself, could not drive nanotechnology development:

> Most foreseeable applications are still 10 or 20 years away from a commercially significant market; however industry generally invests only in developing cost-competitive products in the 3 to 5 year timeframe. It is difficult for industry management to justify to their shareholders the large investments in long-term, fundamental research needed to make nanotechnology-based products possible. Furthermore, the highly interdisciplinary nature of some of the needed research is incompatible with many current corporate structures. (1999b, p. 2)

As Lane and Kalil (2005) saw it, "[G]iven that international leadership in nanotechnology is up for grabs, allowing US funding to stagnate while foreign governments continue to provide double-digit increases seems to us to be an incredibly risky strategy" (p. 52). These sentiments were echoed in President Clinton's (2000) Caltech speech launching the NNI, when he justified significant public investment in the NNI on the grounds that "some of these research goals will take 20 or more years to achieve. But that is why . . . there is such a critical role for the federal government." The potential returns were predicted to be substantial: at the time the NNI was launched, Roco and other experts estimated a $1 trillion market for nanoproducts and a demand for 2 million workers worldwide by 2015 (NSTC, 2000; Roco & Bainbridge, 2001).

Roco and his colleagues proceeded to organize industry support, soliciting endorsement letters that were eventually coordinated through a website set up in 2000 (NSF, 2009).[12] As Lane later recalled in an interview,

> There certainly was interest from industry but there wasn't a groundswell in my recollection. We're talking about the late 90's. I think only a few companies were involved with nano; only companies like IBM or HP had the resources to play around, if you like, with nanotechnology. The sorts of grand challenges that we talked about were really for new breakthroughs in materials, electronics, computing, medicine, etc. This was more of a vision that many scientists, including myself, shared . . . I think there is no question that in terms of numbers of investigators, the larger numbers were in the universities and to some extent the national labs, e.g. NASA and DOE. (N. Lane, personal communication, August 20, 2009)

Since its inception in FY 2001, the NNI has invested more than $18 billion in nanoscale science and engineering (including the 2013 budget request of $1.8 billion). Appropriations increased rapidly initially and have continued growing steadily, albeit more slowly in recent years. Although 15 federal agencies participate in the NNI, there are 5 key agencies accounting for more than 90 percent of the proposed 2013 budget: $443 million for the Department of Energy (DOE), $409 million for the National Institutes of Health in the Department of Health and Human Services, $435 million for the National Science Foundation, $289 million for the Department of Defense, and $102 million for the National Institute of Standards and Technology in the Department of Commerce (PCAST, 2012). Whereas funding for most of these agencies has been relatively stable for the past several years, the proposed DOE appropriation reflects a 42 percent increase over 2012, due in part to DOE's central role in three "Nanotechnology Signature Initiatives" first introduced in 2012: Solar Energy Collection and Conversion, Sustainable Nanomanufacturing, and Nanoelectronics.[13] The reasoning behind these Initiatives is described in the *Supplement to the President's 2012 Budget* in no uncertain terms: "Consistent with the President's Strategy for American Innovation, the Nanotechnology Signature Initiatives aim to catalyze breakthroughs for national priorities and harness science and technology to address 'grand challenges' of the 21st century" (NNI, 2011, p. 8).

Substantial public investment in nanotechnology, with special focus on the three Signature Initiatives, would clearly seem to reflect an effort to shape the future of industrial development. Does this constitute an industrial policy?

Although most of NNI spending is directed at the research end of the research-development-commercialization continuum, there is also limited funding for commercialization under the Small Business Innovation Research (SBIR) and Small Business Technology Transfer (STTR) programs, which together provide some $2 billion annually, on a competitive basis, as startup funding for promising high-tech firms. NNI's contribution to these two programs—which grew from $97 million in 2008 to $127 million in 2009 (the most recent year for which figures are available), for a cumulative total of $204 million since 2004—is described as "in keeping with the Administration's Startup America program, which is aimed at helping entrepreneurs overcome the 'valley of death' on the way to successful commercialization of science and technology breakthroughs" (NNI, 2011, p. 8). Additionally, federal departments and agencies that receive NNI funding also occasionally make direct grants to private firms. Drawing on the Office of Science and Technical Information database, which collects grant information from nine federal agencies, we extracted nanotechnology-related projects and categorized whether or not the recipient was in the private sector.[14] In recent years, roughly 20 percent of DOE's nano-related grants have gone to the private sector; the corresponding figure for NSF is 7–8 percent.[15] It seems clear that only a small amount of federal funding goes directly to the private sector; nor is there evidence that such support is increasing.

In other words, even though the NNI is intended to promote U.S. competitiveness through the commercialization of nano-enabled products, the lion's share of funding is directed at more basic research, under the assumption that this will eventually pay off in success in the global marketplace. The NNI is remarkable not only because the federal government has made a massive multi-agency investment, but also because it reflects a convergence between science and technology policy and a soft form of industrial policy. The driving vision beyond the creation of the NNI reflected the belief that industries needed science, and that science and industries needed government—a convergence of interests that have resulted in a quasi-industrial policy around nanotechnology. The U.S. government selected the technology, launched the NNI, and became the single largest investor. Whereas the NNI may have started at the top, it incorporated participation from the bottom as the policymaking process matured.

The U.S. government may lack a single, grand, national science and industrial strategy like that of China, whose promotion of nanotechnology to be a national priority is discussed subsequently, where there are single ministries in charge of industrial issues. However, the emergence and growth of the NNI shows that federal policy was far more integrated than one might have thought. The integration of science and technology policy and industrial policy with the NNI came with a peculiar embrace of two contrasting ideologies: a strong role for government to invest in science and and an equally strong role for the market economy. Although the goals of the NNI went beyond a concern with developing science and technology to embrace industrial policy concerns with competitiveness, the administrative structure and mechanisms required to promote commercialization are largely absent. This somewhat schizophrenic approach no doubt reflects the fact that although U.S. economic policy is widely seen as market-rational, in practice it is closer to the market-ideological end of the spectrum, requiring what Block (2011) has termed "the invisible hand of government" to indirectly provide some degree of industrial guidance (rather than a more open approach to industrial policy).

SETTING NANOTECHNOLOGY AS A RESEARCH PRIORITY IN CHINA

If the United States falls toward the ideological end of the market-rational/market-ideological spectrum, China can be said to be somewhere between plan-ideological and plan-rational. China has had a long history of using centralized state planning to promote the development of economic growth and social development. The legacy of planning in science and technology was initially because of Soviet influence in the 1950s, and second, because of China's successful completion of its centrally planned strategic weapons program in the 1960s that further strengthened its belief in the planning mechanism.[16] Such efforts have been characterized by prioritizing the areas that could boost China's international status or otherwise satisfy politically motivated missions, and then mobilizing the financial, material, and human resources required to achieve them. Of course, planning worked for the strategic weapons program; although new to China, they were already established technologies, suggesting that China's efforts could eventually prove successful. Central planning becomes more problematic when the outcome of scientific research is largely unknown and where the ultimate objective is to achieve global commercial competitiveness. Whereas the selection of nanotechnology as a national priority in science was the result of a process of rational debate and discussion, it also reflected the lingering ideological legacy of centralized state planning—the issuing of 5- and 15-year plans, supervised by central planning agencies and ultimately the Communist Party leadership.

China's interest in prioritizing nanotechnology began when some key Chinese scientists perceived that it had become a "hot" topic internationally. In September 1985, Bai Chunli, a 32-year-old Chinese chemist with a

PhD from the Institute of Chemistry of the Chinese Academy of Sciences (CAS), landed a postdoctoral fellowship at Caltech (where Clinton would launch the NNI several years later), therefore becoming the first Chinese scientist working at the prestigious Jet Propulsion Laboratory (JPL) since 1950.[17] Two years later, Bai returned to China to start a research program around scanning tunneling microscopy (STM) that he had learned at JPL. Bai was among the first Chinese scientists who were exposed to advancements in nanotechnology, and his experience at JPL proved to be crucial not only for his personal career, but also to the development of nanotechnology in China.

Nanoscale research in China dates back at least the mid-1980s when the CAS and the National Natural Science Foundation of China (NSFC) started to support research on the making of STMs, and more generally scientific issues at the nanoscale. The State Science and Technology Commission, the predecessor of the Ministry of Science and Technology (MOST), initiated a 10-year Climbing Program to support research on nanomaterials, which was followed by the National Basic Research and Development Program (known as the 973 Program) that concentrated its support to nanomaterials and nanostructures. In the meantime, the National High-Tech Research and Development Program (known as the 863 Program[18]) also supported research on the application of nanomaterials (Bai, 2001). Chinese scientists did not necessarily foresee the value and significance of the technology—either in terms of pure science or its economic potential in terms of commercial applications. Indeed, China did not fully embrace nanotechnology until countries such as the United States had formulated national nanotechnology initiatives, which made it easier for Bai and his colleagues to make their case to the scientific and political leadership. In the meantime, the overseas members of the Chinese scientific community—those Chinese-origin scientists who stayed abroad after their study rather than returning to China like Bai—helped China move toward the frontier of international nanotechnology research by drawing the attention of Chinese colleagues to advances that were being made elsewhere. According to Xie Sishen (2007), one of China's leading nanoscientists,

> governments around the world and delegations from other countries, especially those from advanced countries, frequently mentioned nanotechnology ... [Their] exchanges and collaborations ... provided information continuously, which made the government realize its importance from pure basic research to application to impacts on the economy and society.

All this set the tone for raising the profile of nanotechnology in China. In late 2000—at roughly the same time that Roco, Lane, and others were arguing for a national nanotechnology initiative in the United States—Bai took the opportunity to lecture to members of State Council's leading group in science, technology, and education, including a couple of members of the Politburo of the CCP Central Committee, leading a group of Chinese experts in proposing that China "should accelerate the industrialization of the nanotechnology and occupy this world-wide frontier area as soon as possible" (NIBC, 2006). At that time, Bai had been vice president of the CAS, the premium research establishment in China's enormous research and development (R&D) system, for four years.[19] More importantly, he also had been an alternate member of the Communist Party of China (CCP) Central Committee, China's *de facto* governing body, since 1997.

Because of his status in Chinese science and his access to China's political leadership, Bai's suggestion was seriously taken and nanotechnology was quickly selected as a research priority (NIBC, 2006). This is reflected in the speech that Chinese President (and CCP Central Committee General Secretary) Jiang Zemin gave the following year to an international forum on nanomaterials, in which he stated explicitly that

> the development of nanotechnology and new materials should be regarded as an important task of the development and innovation in S&T. The development and application of nanomaterials and nanotechnology is of strategic significance to the development of high technology and national economy in China. (NIBC, 2006)

With political support, China's science- and technology-related government agencies immediately acted to develop a national nanotechnology development strategy. In 2001, MOST, with its mandate for the country's S&T-related matters, quickly convened the State Planning Commission (the predecessor of the National Development and Reform Commission, NDRC), the Ministry of Education (MOE), the NSFC, and the CAS to analyze the strengths, weaknesses, opportunities, and threats in the development of nanotechnology in China. This led to the establishment of a national steering committee on nanotechnology in the same year, which in turn formulated an Outline for the National Nano Science and Technology Development (2001–2010) as a roadmap.[20]

Under the guidance and coordination of the national steering committee, chaired by the Minister of Science and Technology and through its chief scientist (Bai Chunli) various nanotechnology-related programs have been supported and implemented, along with a division of labor among key players. MOST, for example, the 863 Program, and the 973 Program are to fund mission-oriented projects.[21] The CAS positions itself in the national nanotechnology landscape with its forward-looking and strategic advantage and strong research capability. Universities have the responsibility for not only conducting cutting-edge research themselves but also for turning out capable nanoscientists. The NSFC awards grants to the best projects and researchers with the possibility of achieving breakthroughs at the frontier of international research, mainly on the basis of scientific merit judged by peer review. Most noticeable is the involvement of the NDRC, which carries a mission similar to Japan's Ministry of International Trade and Industry (MITI) before its role was taken over by Japan's newly created Ministry of Economy, Trade and Industry (METI) in 2001. With NDRC's approval, China has seen the establishment of the National Center for Nanoscience and Technology (NCNST) in Beijing, the National Engineering Research Center for Nanotechnology (NERCN) in Shanghai, and the Nanotechnology Industrialization Base of China (NIBC) in Tianjin. Except for the scientific research-oriented NCNST, the other two organizations were set up for their prospects in commercializing research coming out of places such as NCNST.

The attention from the central government also has led to action at the institutional and local levels. For example, the CAS, along with the Jiangsu provincial government, the Suzhou municipal government, and Suzhou Industrial Park, set up a Suzhou Institute of Nano-Tech and Nano-Bionics (SINANO), in 2006 to carry out "the fundamental, strategic, prospective research in relative fields, aiming at the internationally technological advancement, national strategic demand and future industrial development" (SINANO, 2011). In contrast to the nanomaterials-focused NCNST in Beijing, SINANO is device-, or application- and commercialization-oriented, which explains why local governments and an industrial park would lend their support to the establishment of an academic institution.

Stimulated by the central government's initiative, local governments also have enthusiastically embraced nanotechnology, as exemplified by the Shanghai and Tianjin organizations previously mentioned. There are other institution-building activities as well. For example Zhejiang province, which neighbors Shanghai, hoped to promote Zhejiang University, a national elite university of importance to local economic and social development, as a major player by partnering with UCLA in the United States to set up the Zhejiang–California International Nanosystems Institute, although the results have been mixed (Cheng, 2007).

But of the various local government efforts, "Nanopolis," a nanotechnology-specific science park within the Suzhou Industrial Park, stands as a useful example. Located in the Yangtze River Delta, Suzhou, already a manufacturing powerhouse strongly oriented toward innovation, sees emerging technologies (including nanotechnology) as its new growth engine. Nanopolis is China's first international innovation park that focuses on both nanotechnology and biotechnology. Established in October 2005 with an RMB 550 (US$86 million) million investment from several public entities (the China–Singapore SIP Ventures, SIP S&T Development Corporation, and the SIP Education Development Investment Corporation), it operates as a corporation and has sought to construct an industrial ecology system through seamlessly integrating industrial planning, policy support, resources concentration, platform support, investment, program development, industry-academic collaboration, financial services, and information sharing. It incorporates nanotechnology innovation with engineering, medium-scale trial, and small-scale production, and provides start-ups with services such as subsidies for housing and office space, help with grant applications, subsidies to support patent applications, and one-stop services required to register a company (for example, logistics, environment, workplace and fire safety, and ICT). Because of its dedication, Suzhou has not only attracted the CAS to set up SINANO, but also has become a prime destination for nanotechnology entrepreneurs with its pro-entrepreneurial environment, including services provided to help set up a business.

A somewhat different local effort is seen in the Shanghai Nanotechnology Promotion Center (SNPC), subordinate to the Shanghai Science and Technology Commission, which was founded in July 2000 (Li & Wang, 2006). SNPC is responsible for coordinating efforts across seven university campuses and other institutions, and nine private enterprises. It also provides training for scientists and engineers on the specialized instruments used in nanoscale research and has several university-affiliated "industrialization bases" for the purpose of transferring research on nanomaterials and nanoparticles to the estimated 100–200 small and medium enterprises (SMEs) under its incubation that are reportedly engaged in nano-related R&D in the Shanghai area. Roughly a third of its 25-person staff are science and engineering professionals. It has reportedly spent RMB1 billion (US$126 million) over the past five years, trained 1,500 scientists and engineers on the use of specialized equipment and machinery, and published an introductory textbook on nanotechnology.

There were additional factors that contributed to the selection of nanotechnology as one of the priority areas for support in China. Although the support of China's political leadership for nanotechnology was bolstered by the push from leading scientists both inside and outside of China, the technology itself met one of the themes underlining China's S&T policy in the reform era. With its rapid economic growth and huge foreign reserves generated by its export-oriented industrialization, China can afford to make increasing investments in science and technology. Yet China's financial and human resources are not unlimited. Therefore, as former Chinese President (1993–2003) Jiang Zemin put it in 1997, China should "do what it needs and attempt nothing where it does not" (*you suo wei, you suo bu wei*),[22] being selective in supporting research endeavors and concentrating its growing but scarce resources where a high payoff is deemed most likely. That is, only those areas whose breakthroughs may not only significantly change the scientific landscape but also bring about huge economic benefit could stand out. Nanotechnology seems to have emerged as a winner, because it holds promise for solving key challenges facing China in agriculture, the environment, population, health, and national security. Choosing nanotechnology has symbolic significance as well: It signifies China is embracing an apparently cutting-edge aspect of global S&T development, something that promises to leverage China's growing scientific advantages.

Finally, the selection of nanotechnology also reflects an ambitious goal of the Chinese state—leapfrogging to the next technology frontier by building up an indigenous innovation capability. Although the reform of China's S&T management system since the mid-1980s has boosted the emergence of a national innovation system, China's scientific and political leadership has never been satisfied with the "me-too approach" that had long characterized its research endeavors. Various government programs, especially the more basic science-oriented 973 Program, had long tried to reverse the trend. However, China's export-led industrial growth, although still regarded as crucial, came to be viewed as insufficient by itself: Not only did the largest profits remain with the multinationals that were producing in China, but the degree of technology transfer—especially with the most advanced technologies—was also limited. China's leaders came to view China as too dependent on foreign technology.

Therefore, in the early twenty-first century, a strategy focusing on talent, patents, and technical standards was formulated by the MOST, promoting a building-up of what was termed "indigenous innovation" capability. This eventually led to the release of the 15-year Medium and Long-Term Plan for the Development of Science and Technology (2006–2020), in early 2006. The MLP (as it is generally known in English) called for China to become an "innovation-oriented nation" by 2020 and a world leader in science and technology by 2050. To accomplish this, it identified four key science areas (so-called "megaprograms") for significant investment, one of which is nanotechnology, on the grounds that they have the potential to contribute significantly to innovative breakthroughs.[23] The MLP sees its emphasis on indigenous innovation as enabling China to achieve "technological leapfrogging"—moving directly into high-impact emerging technologies, rather than merely being stuck in labor- and resources-intensive, low-wage, and low- to intermediate-technology exports, thereby bypassing the more traditional step-by-step movement up the value chain. Indeed, China wants to harness its human capital and promote indigenous innovation in order to address its social, environmental, global competitive, and national security challenges (Cao, Suttmeier, & Simon, 2006).[24] To support the indigenous innovation efforts, various government agencies also worked out a portfolio of innovation policies, which, according to the current understanding of China's innovation policymakers, include science and technology policy, industrial policy, fiscal policy, tax policy, and financial policy (Liu, Simon, Sun, & Cao, 2011).

The MLP formulation actually followed the planning mentality and legacy of the strategic weapons program, by identifying a number of "winners" in the name of "megaprograms" in science and engineering. These programs are supposed to receive investment in the scale of tens if not hundreds of billions of RMB (millions to tens of millions of dollars) over the plan period from the government, with matching funds from other sources, including enterprises. Although nanotechnology eventually emerged as one of the four science megaprograms, it did not initially make the short list. Although we have not yet been able to ascertain why nanotechnology emerged as one of the final four, we imagine that it must have had something to do with the influence of Bai Chunli. Indeed, few among Chinese scientists would have access to the top political leadership; and indeed, with his alternate membership in the CCP Central Committee, he himself is part of the leadership.

Examining nanotechnology's emergence as a top science area from an S&T policy perspective, it would seem that China is now facing a dilemma, reflected paradoxically in its selection of nanotechnology as one of the MLP's science megaprograms. The decision was made apparently because of the tremendous potential that nanotechnology promises to bring—trillions of dollars globally in new products, according to some predictions. China clearly does not want to miss this opportunity. However, by including nanotechnology as a science megaprogram—as opposed to an engineering megaprogram—the MLP recognizes that nanotechnology remains a basic science, with its trillions of dollars in commercial benefits years in the future (if indeed they materialize at all).[25]

In other words, although it is both visionary and appropriate for China (and other countries) to invest in nanotechnology, given its interdisciplinary nature and commercial prospects that are years, if not decades, away, nanotechnology was unlikely to be part of MLP's engineering mega-programs. As Shi Yili, a nanoscientist at Shanghai University, candidly indicated to us, nanotechnology is just an enabling technology, nothing more and nothing less (Shi, personal communication, August 28, 2009). The only possibility for nanotechnology to be picked up as a winner was to promote it as a basic science discipline while using its potential for contributing to economic growth as the attraction. And the strategy worked—nanotechnology, which originally was not a candidate for the science mega-programs, won the support of the political leadership in the final deliberation. Nanotechnology made the cut as part of China's planning for industrial development—acknowledged because of its potential as basic science, yet chosen (and funded at the state, local, and provincial levels) because of its potential commercial payoff.

CONCLUSION

Both the United States and China have invested heavily in nanotechnology. Both have done so because of nanotechnology's perceived potential for triggering a future economic bonanza. In both cases, however, it is recognized that nanotechnology is at best an enabling technology still in its infancy, close to the basic research end of the research-development-commercialization continuum.

In China, as a consequence, nanotechnology's inclusion in central planning schemes such as the MLP required it to be classified as a science—even though in practice it is often funded as if it were an engineering technology close to yielding commercially viable products. In the United States, even though nanotechnology was sold to Congress because of its commercial potential, its public funding continues to emphasize its role as a science in the basic research stage. U.S. federal funding for nanotechnology commercialization remains limited to a handful of small government programs.

China is steering a middle course between its legacy of central planning and its move to more market-driven economic growth. According to recent OECD (2009) estimates, state-owned enterprises (SOEs) still account for a substantial part (roughly a third) of China's industrial output. Critics of China's state-led development (such as the U.S. Chamber of Commerce) claim that SOEs are the favored—and often inefficient—recipients of the state's high-tech investments (McGregor, 2010).[26] Our research, which has focused thus far primarily on private firms, suggests that China's efforts to achieve "indigenous innovation" in such high-tech areas as nanotechnology involve substantial state funding at all levels (central, provincial, local), from basic research to final product. The more local the funding, the more likely it is to cross the "valley of death" in hopes of pushing products to market. In other words, the Chinese government is engaged in what Johnson (1982) termed plan-rational industrial policy, as well as playing a facilitative role in providing the infrastructure, science parks, and greenfield university campuses that may eventually let a thousand nano-based products bloom.

The U.S. government is playing a less aggressive role, consistent with its antipathy toward anything overtly representing industrial policy. Having picked nanotechnology as a winner in 2000, the federal government has since invested heavily in basic nanoscale research, but has been far more timid in picking winning industries, much less particular firms. The NNI was not the result of market-driven imperatives, but the entrepreneurial efforts of Washington insiders, whose vision was then expanded to include a widening circle of supports, including representatives of industries that would hopefully reap its benefits. As was the case in China, the adoption of nanotechnology as a candidate for government funding was largely a top-down effort. Unlike China, however, there was no central planning apparatus to translate that highly successful effort into a sustained effort to promote industrial development. The U.S. effort might be thought of as a stealth industrial policy, brought into the house—at least part way—through the back door. Its funding has been largely intended to facilitate rather than drive its development—O'Riain's (2004) "horizontal technology policy," Block's (2008) "hidden developmental state." Given the current fiscal resources of the U.S. government, particularly in comparison with China, its ability to play a strongly facilitative role is at present also limited.

It is clearly too soon to see which approach will pay off, because the major returns to investment in nanotechnology still lie in the future. The United States has one significant advantage over China—its scientific culture remains far more innovative (Appelbaum & Parker, 2012). Yet even here China is making advances, and—if it even partly succeeds in becoming an "innovative society" through government planning and focused spending—it may also succeed in breathing some life into the concept of industrial policy.

ACKNOWLEDGMENTS

This material is based upon work supported by the National Science Foundation under Cooperative Agreements Nos. SES 0531184 and #SES 0938099 to the Center for Nanotechnology in Society at University of California at Santa Barbara. Any opinions, findings, and conclusions or recommendations expressed in this material are those of the authors and do not necessarily reflect the views of the National Science Foundation.

NOTES

1. The distinction between plan- and market-rationality actually originated with Dahrendorf (1968), who developed it to distinguish between capitalist and state socialist economies. Johnson built on Dahrendorf's distinction.
2. For an excellent discussion of these points on which the present discussion is partly based, see Whitford and Shrank (2011).
3. For a detailed discussion of the emergence of the NNI on which this section is partly based, see Motoyama, Appelbaum, and Parker (2011).
4. See, for example, Lécuyer's (2006) analysis of the rise of Silicon Valley between 1930 and 1970, in which military procurement helped to shape technological innovation and industrial growth.
5. Reich also hosted a four-part public television series, *Made in America*, calling for a more activist government role to enable the United States to compete with a rising Europe. Reich was effectively muzzled during his brief tenure as Labor Secretary, despite his personal relationship with President Clinton (they had been friends since their early twenties, when both were Rhodes Scholars at Oxford); talk of industrial policy had effectively become anathema after the Reagan presidency. Reich (2009) came to regard U.S. efforts in this area as haphazard, responding mainly to industrial crises.
6. Roco is a mechanical engineer who had been at the NSF since 1990 and was serving at the time (1999) as a program director; he is now the NSF's Senior Advisor for Nanotechnology, and has emerged as the chief spokesperson and *de facto* head of the NNI.
7. Hu is currently the Gordon McKay Professor of Applied Physics and Electrical Engineering at Harvard University.
8. Siegel is currently the Robert W. Hunt Professor of Materials Science and Engineering and Director of the Rensselaer Nanotechnology Center at Rensselaer Polytechnic Institute.

9. Lane served as NSF Director from October 1993 to August 1998, then as Assistant to the President for Science and Technology and Director of OSTP in the White House from August 1998 to January 2001. He has since returned to Rice University, where he is the Malcolm Gillis University Professor.
10. Kalil has returned to the White House, where he serves in the Obama Administration as Deputy Director for Policy at the Office of Science & Technology Policy (OSTP); previously he was Special Assistant to the Chancellor for Science and Technology at the University of California, Berkeley.
11. PCAST is a nongovernmental, nonstanding advisory group that does not require a congressional approval. Each president set it up under his own initiative and Clinton established it in 1993 by an executive order.
12. At the NSF website, it says, "If you are interested in endorsing this initiative, please contact M.C. Roco, e-mail: mroco@nsf.gov, Chair NSET, NSTC/WH." Then, the page lists endorsing messages from industry and academic experts. Available from http://www.nsf.gov/crssprgm/nano/reports/endorse.jsp (accessed 6/13/12).
13. There are additional reasons for this increase, including "the maturation of some aspects of nanoscale science and technology that have received long-term, sustained support from the Office of Science and other NNI basic research agencies, to the point where these areas are ripe for the development of nanotechnology enabled applications to address the nation's energy needs" (NNI, 2011, p. 9).
14. The Federal R&D Project Summaries was incomplete: some agencies do not provide data, and although Department of Defense data was included, we excluded it from our analysis because recipient details were lacking due to security concerns. The search from this database did not produce any results from National Institute for Health or Small Business Administration. For DOE and NSF, we used databases managed by each agency because they produced higher search results.
15. The number of grants is not a perfect indicator of the amount of funding, however. We were able to estimate direct private sector NSF-NNI funding, which averaged only 4.2 percent of all NSF-NNI funding during the period 2001–2008.
16. China exploded its first atomic bomb in 1964, launched its first nuclear missile in 1966, exploded its first hydrogen bomb in 1967, and launched its first satellite in 1969.
17. Qian Xuesen (known as H. S. Tsien in the United States) was one of JPL's founders. Stripped of his security clearance in 1950 during the McCarthy period, he spent five years under house arrest, before being repatriated to China in exchange for US pilots captured during the Korean War. Qian returned to China, where he became known as the father of China's space program (Chang, 1996).
18. The numbers that identify China's programs refer to the dates in which they were enacted. Thus, the 863 Program was enacted in March 1986; the 973 Program was enacted in March 1997.
19. He was promoted to CAS executive vice president with the full-minister rank in 2004, and is now its president.
20. A new national steering committee was formed in June 2007 and Bai Chunli still is chief scientist.
21. These programs, initiated in 1986 and 1997, thus considerably predate China's formal move into nanotechnology. When the Medium and Long-Term Plan (MLP) for the Development of Science and Technology

(2006–2020)—also administered by the MOST—was adopted in 2006, the 863 and 973 Programs were placed under its aegis. As we mention below, the MLP identified nanotechnology as one of four "science megaprograms" slated for funding.

22. This theme was taken from Jiang's report to the 15th CCP National Congress in 1997, which reads, "We should formulate a long-term plan for the development of science from the needs of long-range development of the country, taking a panoramic view of the situation, emphasizing key points, *doing what we need and attempting nothing where we do not*, strengthening fundamental research, and accelerating the transformation of achievements from high-tech research into industrialization" (emphasis added). This was in turn adapted from the May 1995 decision of the CCP and the State Council to push forward China's S&T progress, although the wording was slight different—"catching up what we need and attempting nothing where we do not" (*you suo gan, you suo bu gan*).
23. The other three science megaprograms are quantum research, reproductive and developmental biology, and protein science. The MLP also identified 16 engineering megaprograms, eight frontier technologies, and 11 key areas for funding. A detailed discussion of these is found in Cao, Suttmeier, and Simon (2006).
24. For a more detailed discussion of the MLP and the reaction of U.S. businesses (as well as the U.S. government) to China's emphasis on indigenous innovation, see Chapter 8 in this volume.
25. The establishment of SINANO within CAS also may reflect such a dilemma. Although CAS as a whole is supposed to be engaged in basic research, SINANO, given its device and application orientation, is located in the downstream of the R&D spectrum.
26. The Chamber report, which derides the MLP as "a web of industrial policies" (the report's subtitle), views indigenous innovation as a movement away from market liberalization in favor of "hunkering behind the 'techno-nationalism' moat," switching "from defense to offense" in light of China's economic ascendance as well as its fear of foreign domination (McGregor, 2010, pp. 6–7). For a more detailed discussion see Appelbaum and Parker (2012).

REFERENCES

Appelbaum, R. P., & Henderson, J. W., (Eds.). (1992). *States and development in the Asian Pacific Rim*. Newbury Park, CA: Sage Publications.

Appelbaum, R. P., & Parker, R. (2012). The Chinese century? Some policy implications of China's move to high-tech innovation. In B. Harthorn & J. Mohr (Eds.), *The social life of nanotechnology*. New York: Routledge.

Bai, C. (2001). *Nano science and technology research in China*. Retrieved September 2009, from http://www.channelwest.com/FILES/westqikan/westqikan2/03.htm (in Chinese)

Block, F. (2008a). America's stealth industrial policy. *Miller-McCune*. Retrieved August 2009, from http://www.miller-mccune.com/business-economics/americas-stealth-industrial-policy-4516/

Block, F. (2008b). Swimming against the current: The rise of a hidden developmental state in the United States. *Politics and Society, 36*, 169–206.

Block, F. (2011). Innovation and the invisible hand of government. In F. Block & M. R. Keller (Eds.), *State of innovation: The U.S. government's role in technology development*, pp. 1–16. Boulder, CO: Paradigm.

Breznitz, D. (2007). Industrial R&D as a national policy: Horizontal technology policies and industry-state co-evolution in the growth of the Israeli software industry. *Research Policy, 36,* 1465–1482.

Cao, C., Suttmeier, R. P., & Simon, D. F. (2006). China's 15-year science and technology plan. *Physics Today, December,* 38–43.

Castells, M. (Ed.). (2004). *The network society: A cross-cultural perspective.* Cheltenham, UK: Edward Elgar.

Chang, H. (1994). *The political economy of industrial policy.* New York: St. Martin's Press.

Chang, I. (1995). *Thread of the silkworm.* New York: Basic Books.

Cheng, J. 2007. Interview conducted at the Zhejiang University, Zhejiang-California International Nanosystems Institute, August 1, 2007.

Cientifica. (2011). Global nanotechnology funding 2011 (July 13). Retrieved March 2012, from http://www.azonano.com/news.aspx?newsID=22980

Clinton, W. J. (2000). *Remarks by the President, at science and technology event.* Los Angeles, CA: Office of the Press Secretary, the White House.

Dahrendorf, R. (1968). Market and plan: Two types of rationality. In R. Dahrendorf (Ed.), *Essays in the theory of society*, pp. 215–231. London: Routledge and Kegan Paul.

Evans, L. (2004, June 2). *Government's role in development: The case of Brazil under the Worker's Party.* Retrieved June 2004, from http://www.international.ucla.edu/article.asp?parentid=11691

Graham, O. L. (1992). *Losing time: the industrial policy debate.* Cambridge, MA: Harvard University Press.

Hall, B. H., & Rosenberg, N. (2010). *Handbook of the economics of innovation* (Vol. 2). Oxford: North-Holland.

Hall, P. A., & Soskice, D. W. (2001). *Varieties of capitalism: The institutional foundations of comparative advantage.* Oxford: Oxford University Press.

Henderson, J., & R. P. Appelbaum. (1992). Situating the state in the East Asian development process. In R. P. Appelbaum & J. W. Henderson, (Eds.), *States and development in the Asian Pacific Rim,*, pp. 1–26. Newbury Park, CA: Sage Publications.

Johnson, C. A. (1982). *MITI and the Japanese miracle: The growth of industrial policy, 1925–1975.* Stanford, CA: Stanford University Press.

Lane, N. (2008). US Science and technology: An uncoordinated system that seems to work. *Technology in Society, 30,* 248–263.

Lane, N., & Kalil, T. (2005). National Nanotechnology Initiative: Present at the creation. *Issues in Science and Technology, Summer,* 49–54.

Lécuyer, C. (2006). *Making silicon valley: Innovation and the growth of high tech, 1930–1970, inside technology.* Cambridge, MA: MIT Press.

Li, N., & Wang, K. (2006). Interview conducted at breakfast meeting of the American Chamber of Commerce, Beijing. August 8, 2007.

Liu, F., Simon, D. F., Sun, Y., & Cao, C. (2011). China's innovation policies: Evolution, institutional structure, and trajectory. *Research Policy, 40,* 917–931.

McCray, W. P. (2005). Will small be beautiful? Making policies for our nanotech future. *History and Technology, 21*(2), 177–203.

McCray, W. P. (2006). Thomas Kalil: Transcript of an oral history interview. Santa Barbara, CA: Center for Nanotechnology in Society, University of California, Santa Barbara.

McGregor, J. (2010). China's drive for indigenous innovation: A web of industrial policies. *U.S. Chamber of Commerce Global Regulatory Cooperation Project.* Retrieved May 2010, from http://www.uschamber.com/sites/default/files/reports/100728chinareport_0.pdf

Motoyama, Y., Appelbaum, R., & Parker, R. (2011). The National Nanotechnology Initiative: Federal support for science and technology, or hidden industrial policy? *Technology in Society, 33*(12), 109–118.

National Nanotechnology Initiative (NNI). (2011, February). *Supplement to the President's 2012 budget.* Retrieved February 2011, from http://www.nano.gov/sites/default/files/pub_resource/nni_2012_budget_supplement.pdf and Frequently Asked Questions. Retrieved from http://www.nano.gov/nanotech-101/nanotechnology-facts

NIBC. (2006). Brochure provided by the Nanotechnology Industrialization Base, China, during interview, August 6, 2006.

National Science Foundation (NSF). (2009, February 25). *NNI Endorsement,* edited by National Science Foundation. Retrieved February 2009, from http://www.nsf.gov/crssprgm/nano/reports/endorse.jsp

National Science and Technology Council (NSTC). (2000). *The National Nanotechnology Initiative: The initiative and its implementation plan.* Washington DC: National Science and Technology Council.

Organisation for Economic Co-operation and Development (OECD). (2009, January 26). *State-owned enterprises in China: Reviewing the evidence.* OECD Working Group on Privatisation and Corporate Governance of State Owned Assets Occasional Paper. Retrieved January 2009, from http://www.google.com/url?sa=t&rct=j&q=&esrc=s&source=web&cd=1&cts=1331582947920&ved=0CCYQFjAA&url=http%3A%2F%2Fwww.oecd.org%2Fdataoecd%2F14%2F30%2F42095493.pdf&ei=3VdeT46-MIKNigKOs6WpBA&usg=AFQjCNG7CDTVfLEuCpvUxxtVF6jG8m06uw&sig2=iB6fQCIjoGaUs5ks6RwWFA

O'Riain, S. (2004). *The politics of high-tech growth: Developmental network states in the global economy, structural analysis in the social sciences.* New York: Cambridge University Press.

Pempel, T. J. (1999). *The politics of the Asian economic crisis.* Ithaca, NY: Cornell University Press.

Powell, W. (1990). Neither market nor hierarchy: Network forms of organization. In B. Staw & L. L. Cummings (Eds.), *Research in organizational behavior,* pp. 295–336. Greenwich, CT: JAI Press.

President's Council of Advisors on Science and Technology (PCAST). (1999a). *Letter to the President endorsing a National Nanotechnology Initiative.* Washington, DC: President's Advisors on Science and Technology.

President's Council of Advisors on Science and Technology (PCAST). (1999b). *Review of proposed National Nanotechnology Initiative.* Washington, DC: President's Committee of Advisors on Science and Technology, Panel on Nanotechnology.

President's Council of Advisors on Science and Technology (PCAST). (2012). *Report to the President and Congress on the Fourth Assessment of the National Nanotechnology Initiative.* Washington, DC: President's Committee of Advisors on Science and Technology, Panel on Nanotechnology."

Reich, R. B. (1997). *Locked in the cabinet.* New York: Vintage.

Reich, R. B. (2009). What industrial policy should be. *American Prospect.* Retrieved March 2012, from http://prospect.org/article/what-industrial-policy-should-be

Roco, M. C. (2007). National Nanotechnology Initiative—Past, present, future. In W. M. Goddard III, D. W. Brenner, S. E. Lyshevski, & G. J. Iafrate (Eds.), *Handbook of nanoscience and engineering* (2nd ed.), 3.1–3.26. Boca Raton, FL: Taylor and Francis (CRC Press).

Roco, M. C., & Bainbridge, W. S. (2001). *Societal implications of nanoscience and nanotechnology.* Boston: Springer.

Rodrik, D. (2004). *Industrial policy for the twenty-first century* (KSG Working Paper No. RWP-04–047). Cambridge, MA: Harvard University Press.

SINANO. (2011). Brief introduction. Retrieved December 15, 2011, from http://english.sinano.cas.cn/au/bi/

Teubal, M. (1997). A catalytic and evolutionary approach to horizontal technology policies (HTPs). *Research Policy, 25*, 1161–1188.

Wade, R. (1990). Industrial policy in East Asia: Does it lead or follow the market? In G. Gereffi & D. L. Wyman (Eds.), *Manufacturing miracles: Paths of industrialization in Latin America and East Asia*, pp. 213–231. Princeton, NJ: Princeton University Press.

Whitford, J., & Shrank, A. (2011). The paradox of the weak state revisited. In F. Block & M. R. Keller (Eds.), *State of innovation: The U.S. government's role in technology development*, pp. 261–281. Boulder, CO: Paradigm.

Woo-Cumings, M. (1999). *The developmental state.* Ithaca, NY: Cornell University Press.

WTEC. (1998). R&D status and trends in nanoparticles, nanostructured materials, and nanodevices in the United States. *Proceedings of the May Workshop.* Baltimore, MD: International Research Institute at Loyola College in Maryland. Retrieved December 15, 2011, from: www.wtec.org/pdf/nanousws.pdf

WTEC. (1999). *Nanostructure science and technology: A worldwide study.* Baltimore, MD: WTEC, Loyola College. Retrieved December 15, 2011, from: http://www.wtec.org/loyola/nano/toc.htm

Xie, S. (2007). *Funding and networks for nanotechnology in China.* Presentation given at the Institute for Physics, Beijing, July 17, 2007.

8 The Chinese Century?

China's Move Towards Indigenous Innovation: Some Policy Implications

Rachel Parker and Richard P. Appelbaum

> The rules have changed. In a single generation, revolutions in technology have transformed the way we live, work, and do business . . . Today, just about any company can set up shop, hire workers, and sell their products wherever there's an Internet connection. Meanwhile, nations like China and India realized that with some changes of their own, they could compete in this new world. And so they started educating their children earlier and longer, with greater emphasis on math and science. They're investing in research and new technologies. Just recently, China became the home to the world's largest private solar research facility and the world's fastest computer.

> Our infrastructure used to be the best, but our lead has slipped . . . China is building faster trains and newer airports. Meanwhile, when our own engineers graded our nation's infrastructure, they gave us a D.
>
> —U.S. President Barack Obama, State of the Union Address, January 25, 2011

CHINA: AN EMERGING HIGH-TECH GLOBAL POWER

According to Gabriele, the Chinese state employs an important but not unique tool in its ownership and control rights on its largest and most advanced industrial enterprises. However, thanks to its unique degree of control on the country's resources, it also engages in huge and ever-increasing investments in infrastructure, institution- and human capital building, R&D, and in other areas, on a scale unequalled anywhere else in the world (Gabriele, 2009, 17). This public investment drive generates a network of systemic external economies, which in turn decisively enhance the competitiveness, productivity, and profitability of both public and privately owned/controlled industrial enterprises . . . "we argue that the role of the State (to be understood as a holistic term referring to the public sector as whole), far from being withering out, is in fact massive, dominant, and crucial to China's industrial development" (Gabriele, 2009, pp. 3).

China has been described as "a nation led by technocrats" trained "typically in narrow and specialized fields of science and technology," who view China's future as capable of management by experts (Suttmeier, 2007, pp. 71–72). China's leaders have ample reasons to hold such beliefs: China overtook Japan as the world's second largest economy in 2010. The country's meteoric economic growth has even been heralded as launching a new "Beijing Consensus," a state-driven model for economic growth that effectively challenges the neoliberal "Washington Consensus."[1] Whereas China's growth has largely been the result of market liberalization and export-oriented policies that have proven highly friendly to foreign firms, in recent years there has been a significant shift in approach—one that China's critics see as a return to more heavy-handed state-run economic policies of the past. The critic's chief cause of concern is China's recent decision to compete not only on the basis of low-cost manufactured goods, but instead to become a global high-tech player "at the forefront of world technology development . . . focusing on 'leap-frog' development in key high-tech fields in which China enjoys relative advantages" (MOST, 2005). The ambitious plan to accomplish these goals is laid out in China's *Long and Medium Term Scientific and Technological Development Plan Guidelines for the Period 2006–2020* (hereafter MLP), issued in December 2005. The MLP called for China to invest heavily in research and development in advanced technologies, calling for China to become an "innovation-oriented society" by 2020 and a world leader in science and technology by 2050. The adoption of the MLP was a key component of a broad shift in economic strategy for China: to move away from dependence on exports as the sole driver of economic growth by developing the capacity for high-tech innovation. This, in turn, would enable Chinese firms—state owned and private—to compete at the top of the value chain, where the largest returns are realized.

The MLP also apparently signals an increased emphasis on state planning, for which final say lies with the Chinese Communist Party (CCP). The Party exerts its influence through Leading Groups created within the State Council, which coordinate large-scale planning across government agencies. The Leading Groups are typically chaired by a vice premier or higher level figure who also belongs to the CCP Central Committee Politburo or its Standing Committee—China's *de facto* governing body (Appelbaum, Parker, Cao, & Gereffi, 2011). The State Leading Group for Science, Technology, and Education, which has been led by premier Wen Jiabao, is influential in setting the nation's science, technology, and education policy. As we shall see, this perceived shift to greater state planning, through which China hopes to become a global economic and political player, is strongly criticized by organizations that represent foreign firms, as well as by the U.S. government.

China has placed its bets largely on the creation of a national innovation system that will result in indigenous innovation in leading-edge areas of science and technology, which are seen as key to national prosperity

(NIBC, 2006). Both the 11th and 12th Five-Year Plans (2006–2010, 2011–2015) view innovation as the centerpiece of China's economic strategy, the means to address the country's significant social, environmental, global competitive, and national security challenges. China's plans for "leapfrogging development" bypass the more traditional step-by-step movement up the value chain, requiring state investment in areas where firms are unable or unwilling to (Appelbaum et al., 2011). Although China's leadership does not call for abandoning its export-led industrial growth, it regards this strategy as insufficient by itself: Not only have the largest profits remained with the multinationals that are manufacturing in China, but the degree of technology transfer—especially with the most advanced technologies—has also proven to be limited.

China's development strategy now calls for a three-pronged approach. First, to continue to foster its export sector (a major source of employment, and one in which wages are slowly rising); second, to develop its domestic market, a potentially profitable source for its burgeoning domestic industries; and third, to foster the growth of high-technology development, drawing on its rapidly expanding talent pool of low-cost (relative to other advanced industrial countries) scientists and engineers. Given the importance attached to "strengthening the nation through science, technology, and education," China's science and technology (S&T) policy has become a national development strategy since the mid-1990s. China's vast foreign reserves are to finance these efforts, both through the creation of infrastructure (a national network of high-speed trains and highways, ports, and telecommunications), and through government investment in universities, science parks, and targeted research and development of emerging technologies. By 2008, the Chinese government was investing some 4 billion Renminbi (RMB) (US$585 million) in the two principal programs aimed at developing its high tech capacity,[2] as well as providing additional funding for energy sectors such as biofuels. China's spending on research and development (R&D) grew from 0.6 percent of its gross domestic product (GDP) in 1996 to 1.5 percent in 2010, approaching that of Europe (1.69 percent). Although China still falls behind the United States (2.85 percent), government policy calls for reaching U.S. levels by 2020 (Breakthrough Institute, 2009; Cao, Suttmeier, & Simon, 2006; Grueber & Studt, 2009). China seems poised to achieve such growth: R&D spending grew by a quarter between 2008 and 2009, reaching $25.7 billion, much of which is intended for science and technology projects (Li, 2009). China is expected to bypass Japan in R&D spending by 2011, surpassed only by the United States (Naik, 2009). One industry forecast of R&D spending describes China's R&D spending during the past decade as "history making," because it "exceeds and challenges both the US and Europe in terms of the intellectual property it generates and the financial and infrastructure commitments it continues to make in science and technology endeavors" (Batelle, 2009, p. 24; see also Adams, King, & Na, 2009). The study points to such indicators as China's growing share of research publications in virtually every scientific category, its increased success rate in obtaining patents, and its rapid growth in international research collaboration.

As we have shown elsewhere, the maturation of China's State Intellectual Property Office (SIPO) in itself can be taken as a mixed indicator of the country's success (Appelbaum, Parker, & Cao, 2011; Parker, Ridge, Cao, & Appelbaum, 2010). The rate at which researchers are filing nanotechnology patents to SIPO has grown steadily over the past decade. Reflecting China's WTO commitments, in 2008, the National Intellectual Property Strategy was announced, which states as a goal the intent to foster increased capacity for domestic innovation. In that same year, "more than 800,000 patent applications were filed in China's State Intellectual Property Office, by far the largest number received by any patent office in the world" (Zhang, 2009).[3] As early as 2006, China had an R&D workforce that included some 1.2 million scientists and engineers, awarding more than 19,000 doctorates in those areas, trailing only the United States and Russia (Suttmeier, 2008). Moreover,

> China benefits from its "science diaspora" and the international "brain circulation" which brings scientists working in China into active contact with ethnic Chinese colleagues working in some of the world's leading laboratories and high-technology firms. The result of these trends has been the transformation of the Chinese technical community into one that is younger, more achievement orientated, better compensated, increasingly productive, and much more cosmopolitan and in tune with international trends than ever before. (pp. 2–3)

Apart from Chinese nationalism, the motivations for this shift were both political and economic. The paper that announced the MLP noted that China's reliance on the United States, Japan, and other economically advanced countries for key technologies posed a long-term threat to the country's security. Moreover, although China had clearly benefited from foreign investment in terms of its large and growing foreign reserves, it had not realized significant technology transfer, nor reaped the lion's share of the profits:

> If you want to get to the bottom of indigenous innovation . . . for every Chinese-made DVD player sold, the Chinese manufacturer must pay a large royalty fee to the European or Japanese companies that patented various components of the unit, such as its optical reader. These foreign firms reap substantial profits, but the Chinese take is extremely small—and is shrinking further as energy, labor, and commodity prices rise. Policymakers in Beijing, looking to strengthen China's economy, are no longer satisfied with the country's position as the world's manufacturer. Their solution is to break China's dependence on foreign technology,

moving from a model of "made in China" to one of "innovated in China." (Segal, 2010)

The Chinese government consulted widely before the final version of the MLP was drafted; some 2,000 scientists (including leading ex-pat Chinese working in foreign universities and in the private sector), government officials all played a role. Although the scientists reportedly favored strong peer review in the selection of projects for funding,[4] government officials championed an industrial policy approach in which key sectors would be identified for megaprojects—in the view of the MLP's critics, a slippery slope that heralded a return to failed state-run economic policies of the past (McGregor, 2010). The government officials ultimately won out. The MLP called for China to invest heavily in research and development in advanced technologies, calling for China to become an "innovation-oriented society" by 2020 and a world leader in science and technology by 2050. Given the country's limited resources, however, it also concluded that China should "do what it needs and attempt nothing where it does not" (*you suo wei, you suo bu wei*),[5] concentrating its public investments where a high payoff was deemed most likely.

Four "science megaprojects" (one of which is nanotechnology)[6] were therefore singled out as key areas for funding, along with 13 "engineering megaprojects,"[7] 8 "frontier technology" programs,[8] and 11 "key areas"[9] (Cao et al., 2006, Box 2, p. 43). A Special Projects Office was created within MOST to review proposals, approve funding, and monitor projects. And when global financial markets collapsed in November 2008, China launched its 4 trillion RMB (roughly $600 billion) stimulus package, with "science and technology innovation and industrial structure adjustment" identified as one of its ten investment areas. Four percent of the stimulus package is directly intended to fund R&D, along with innovative projects; some will be channeled directly into the MLP (Valigra, 2009).[10] This will further strengthen its industrial policy approach to economic growth (McGregor, 2010).[11] Estimates for total central government funding of nanotechnology itself vary widely, ranging from as little as $230 million for the five-year period 2000–2004 (Bai, 2005), to $160 million in 2005 alone (Bai & Wang, 2007), to $250 million in that same year (Holman et al., 2006). Although even the highest figures are still considerably less than the United States is publicly investing (as noted previously, $1.6 billion in 2010), China's governmental spending on nanotechnology may not be far off when adjusted for differences in labor and infrastructure costs (nanotechwire.com, 2005).

Overseas Chinese scientists are increasingly returning home, attracted by a combination of national pride, growing opportunities for conducting research, the prospect of launching a financially successful high-tech venture, and lucrative financial and other incentives (LaFraniere, 2010). Some relocate permanently (the so-called "sea turtles"[12]); others hedge their bets by splitting their time between appointments in China and their current home universities abroad ("sea gulls") (Appelbaum & Parker, 2011). The Chinese government is doing its best to encourage this return migration through initiatives such as the "Thousand Talents Program,"[13] launched by the CPC Central Committee in January 2009. As summarized by Vivek Wadwha, Senior Research Associate with the Labor and Worklife Program at Harvard Law School, "they are going after the A-grade players. They are basically doing everything they can. They give you labs. They give you everything you want. They make you feel like a national hero" (quoted in Paddock, 2010).

Such initiatives have resulted in the return of several hundred scientists and engineers from the United States, Japan, Britain, Germany, France, and other advanced industrial economies by September of that year. About half are reportedly working in private sector firms (some in their own businesses), and half in universities and research institute (CAS, 2009a).[14] One 2008 survey of 229 Chinese students at American universities found that only 10 percent planned to remain in the United States permanently; 52 percent believed that China offered better job prospects. A similar survey of 637 Chinese returnees reported that 72 percent believed they were doing better professionally by virtue of having comeback to China (Wadwha, 2010)—notwithstanding the challenges many experience when they begin working in Chinese labs, whose hierarchical structures, typically based on seniority and personal networks may stifle creativity (Suttmeier, 2008; Xiao, 2010).

According to official statistics, after China introduced its High-Tech Certification Management Policy in 2008, which slashed corporate taxes from 25 to 15 percent for companies that passed the high-tech certification requirements, some 20,000 were certified in the next year and a half—although there is evidence that as many of half of these were falsely certified in order to receive the tax breaks (Zhou & Yang, 2010). More anecdotally, China has already established a presence in a number of competitive high-tech industries as creators of products with brand name recognition, rather than merely as manufacturers. At the 2010 Consumer Electronics Show, for example, Zhou Houjian, chairman of the $8 billion Chinese electronics manufacturer Hisense, announced that his company will begin selling its own high-end LCD TVs (designed and made in China) in the United States and Australia (Woyke, 2010). The Chinese company Suntech, a global multinational that is now the world's third largest solar company (and the world's largest producer of silicon photovoltaics), is based in Wuxi, China, and in November 2009 planned to open a manufacturing plant in Arizona ("Frost & Sullivan Recognizes Suntech Power," 2008). Applied Materials—the world's largest supplier of solar-manufacturing equipment—announced in December 2009 that it was opening an R&D center in Xi'an. Mark Pinto, Applied Materials' Chief Technology Officer, will also relocate from Silicon Valley to China (Bourzac, 2009).[15]

The conclusions of one recent study of China's science and technology progress, conducted by researchers at Georgia Tech's Technology Policy and Assessment Center (which routinely tracks different countries' progress through the analysis of models based on a large number of "high tech indicators"), are worth quoting at length:

> All this suggests that China is rapidly heading to rival the United States as the principal driver of the world's economy—a position the USA has held since the end of World War II . . . One might well predict that China will surpass the United States in technology-based competitive capabilities within a decade or two. The image of China as just a low-cost producer of manufactured goods is plain wrong. Other data reflect China's expanding research and development activities . . . As China becomes more proficient at innovation processes. . .linking its burgeoning R&D to commercial enterprise, watch out. And China is increasing attention to management of technology . . . to do just that. (Porter et al., 2009, p. 20)

LEGITIMATE CONCERNS—OR CHINA-BASHING?

China's move toward indigenous innovation has not been well received by Western firms, governments, or academic specialists. Yasheng Huang, International Program Professor in Chinese Economy and Business at MIT's Sloan School of Management, sees China as "now in the midst of one of the most statist periods in its reform era" (Huang, 2010). Huang argues that especially since the 2008 global economic meltdown, when China launched its large economic stimulus package, there has been a strong turn to a self-defeating industrial policy. Huang's research seeks to show that the benefits from China's economic growth were more widely shared during earlier, more liberal periods, than they have since 2003, when the Hu Jintao and Wen Jiabao government took office and ushered in its increasingly statist approach. Most economists would agree: a strong industrial policy is likely to result in placing governmental bets (and hence public monies) on losing industries, rather than following the market signals that are more likely to result in fruitful investment decisions. A recent editorial in *Science* by the life science deans at China's two leading universities (Tsinghua and Peking) was highly critical of targeted funding on China's research culture:[16]

> Although scientific merit may still be the key to the success of smaller research grants, such as those from China's National Natural Science Foundation, it is much less relevant for the megaproject grants from various government funding agencies, which range from tens to hundreds of millions of Chinese yuan . . . For the latter, the key is the application guidelines that are issued each year to specify research areas and projects. Their ostensible purpose is to outline "national needs."

> But the guidelines are often so narrowly described that they leave little doubt that the "needs" are anything but national; instead, the intended recipients are obvious. Committees appointed by bureaucrats in the funding agencies determine these annual guidelines . . . This top-down approach stifles innovation and makes clear to everyone that the connections with bureaucrats and a few powerful scientists are paramount, dictating the entire process of guideline preparation . . . The time for China to build a healthy research culture is now, riding the momentum of increasing funding and a growing strong will to break away from damaging conventions. A simple but important start would be to distribute all of the new funds based on merit, without regard to connections. (Shi & Rao, 2010)

A report from the U.S. Chamber of Commerce also views indigenous innovation as a significant change in direction for China. Whereas in the Chamber's view, China had been moving toward greater market liberalization, it now sees the country—or at least its leaders—as "hunkering behind the 'techno-nationalism' moat," switching "from defense to offense" in light of China's economic ascendance as well as its fear of foreign domination (McGregor, 2010, pp. 6–7). Because the MLP calls not only for China to develop its own technology, but also for the country to assimilate imported technologies, the MLP reportedly "is considered by many international technology companies to be a blueprint for technology theft on a scale the world has never seen before" (p. 4).

Such concerns were not alleviated when MOST, the National Development and Reform Commission (NDRC), and the Ministry of Finance (MOF) issued Circular 618 in November 2009,[17] which listed a wide range of products for which firms that embodied indigenous innovation[18] were to be given preferential treatment in government procurement.[19] According to the U.S. Chamber of Commerce Report (McGregor, 2010), such a "Buy-China Plan" would effectively exclude foreign competition, as few foreign-made products would meet the requirements.[20] Foreign access was also said to be hampered by "compulsory certification and standards requirements," as well as "requirements for the disclosure of technology secrets and other proprietary information that serve to exclude foreign products from major Chinese markets" (p. 22).

The Chamber report also speaks to China's nascent intellectual property regime, which—although modeled after international patent law and formally strong—is viewed as poorly enforced in practice. China's patent law also includes utility model patents, intended to protect poorly defined innovations, and which are subject to limited review by SIPO (China's patent office),[21] as well as design model patents that protect design and other aesthetic considerations. The Chamber of Commerce Report views these patents as "junk patents," filed by private firms and state-owned enterprises largely to satisfy government overseers that their funding is producing results.[22] They are also used, at least in the Chamber's view, "to retaliate against foreign companies which file intellectual property infringement lawsuits offshore that stymie the international expansion plans of Chinese companies . . . These 'junk patents' are proving to be a potent weapon against foreign companies" (McGregor, 2010, pp. 26–27). The Chamber Report concludes darkly:

> [M]egaprojects that uncap gushers of government money for civil servants to spray across a landscape of state companies more familiar with R&R [Rest and Relaxation] than R&D are likely to end in a trail of tears . . . As more details of indigenous innovation plans emerge, American and European politicians are seeing an assault on their core national economic strengths. (p. 37)

The U.S. government is no less concerned with China's turn to indigenous innovation than is the U.S. Chamber of Commerce, and has challenged it publicly as well as in high-level talks between the two countries.[23] In its 2010 Report to Congress, the U.S.–China Economic and Security Review Commission[24] strongly criticized "China's pursuit of policies that rely on trade-distorting government intervention intended to promote China's domestic industries and protect them from international competition" (p. 4). The Commission's conclusions are worth quoting at some length:

> Within the last year, the Chinese government has initiated new industrial policies, such as "indigenous innovation," which have further slowed the pace of economic reform and affected the ability of American companies to operate and compete in China. Such policies have also harmed US exporters and import-sensitive domestic firms. To resolve these trade imbalances, the United States has sought remedial action through the WTO, but the lengthy process has at times done irreparable harm to US companies before relief has been granted. (p. 2)
>
> The Chinese government's relations with foreign investors in China appear to be going through a profound change since Beijing announced its indigenous innovation policy, which explicitly favors domestic companies over foreign firms, particularly in government procurement. The American Chamber of Commerce in China reported in its 2010 annual survey that 31 percent of over 300 member companies polled (up from 28 percent in the 2009 annual survey) said their ability to participate and compete in China's market was impeded by discriminatory government policies and inconsistent legal treatment. . . In fact, some businesses have publicly declared that they gradually are being squeezed out of the Chinese market by government policies that first demand technology transfer in exchange for market access and then favor domestic companies. (p. 20)[25]

Indigenous innovation, in the view of the Commission, is but one source of the enormous U.S. trade deficit with China,[26] described as "a major drag on the US economy;" the deficit is thus seen as the result of "a deliberate economic policy that relies on exports and foreign investment capital to amass a large current account surplus with the United States [which] is loaned back to the United States as part of China's deliberate policy" (p. 3). The problem is exacerbated by an artificially low exchange rate, which the Commission estimates at 20–40 percent.

The Commission also expressed concern over China's military modernization, particularly its ability to conduct offensive air and missile operations, improvements that "have expanded China's ability to operate outside its borders and reach US regional allies, such as Japan, as well as US forces in the region." In addition to providing strong political and financial support for military upgrading, the Chinese government assures "market access to domestic aviation manufacturing firms" (p. 5). China's rising economic power—fueled by industrial policies seen as violating WTO free trade principles—is thus matched by growing military capability, to which "foreign aviation manufacturing firms, such as Boeing and Airbus, are compelled to provide technology and know-how offsets in return for market access" (p. 5).[27] China's semi-official public unveiling of its J-20 Stealth fighter, when images of the airplane appeared on Chinese Internet during a January 2011 visit by U.S. Defense Secretary Robert Gates, raised further concerns.

Yet despite such concerns, foreign businesses depend on China for their supply chains, and, perhaps more significantly, much hoped-for future access to China's growing consumer market. Moreover, they continue to sit at the top of the value chain, accounting for the large majority of Chinese exports and imports. In 2009, for example, more than half of exports were produced in foreign business—a figure that rises to 85 percent for the high-tech sector, leading one study to conclude that "foreign companies' profits have yet to suffer under the impact of indigenous innovation policies" (Kennedy, 2010, p.18). Although concerns are frequently expressed that China will eventually turn on its foreign investors, seizing their technology to become potent competitors, a growing number of multinationals are engaging in strategic partnerships. Examples include (Oster, Shirouzu, & Glader, 2010):

- General Electric is about to enter into a 50–50 partnership with the China Aviation Industry Corporation (AVIC), a consortium that makes both military and civilian aircraft, to produce aircraft electronic systems; AVIC seeks to become a globally competitive company
- General Motors has established a joint venture with the Shanghai Automotive Industry Corporation (SAIC Motors), China's oldest car manufacturer, to manufacture and market SAIC's low-cost Wuling microvan in India—a deal that firmly establishes SAIC in India, which also has a growing motor vehicle market (China is now the world's largest vehicle market). GM now owns 44 percent of Wuling (SAIC owns half), whose sales amount to nearly a sixth of GM's global vehicle total.

- Energy firms such as BP and Royal Dutch Shell have entered into multibillion dollar partnerships with state-owned China National Petroleum Company or its publicly traded subsidiary PetroChina, in hopes of exploiting resources in parts of the world where China has established access.
- Crédit Agricole SA, France's largest retail banking group, is partnering with China's state-owned Citric Securities; in exchange for helping Citric to expand its global operations, Crédit Agricole hopes to increase its footprint in China

WILL CHINA BE A PLAYER IN NANOTECHNOLOGY?

Nanotechnology, the latest high-tech revolution, is predicted to result in sweeping social and economic changes. It is defined by the U.S. National Nanotechnology Initiative (NNI) as "the development and application of materials, devices and systems with fundamentally new properties and functions because of their structures in the range of about 1 to 100 nanometers" (Renn & Roco, 2006).[28] At this scale, particularly at the bottom end (10–20 nanometers), "material structures of the same chemical elements change their mechanical, optical, magnetic and electronic properties, as well as chemical reactivity leading to surprising and unpredicted, or unpredictable, effects. In essence, nanodevices exist in a unique realm, where the properties of matter are governed by a complex combination of classic physics and quantum mechanics" (p. 1).

Because of its wide-reaching and diverse potential across industries, nanotechnology is argued to herald the next great technological revolution, one capable of solving many human problems while generating enormous economic returns (Lieberman, 2005; Roco, Williams, & Alivisatos 1999). As of 2008 at least 60 countries, rich and poor, were collectively investing more than $8.4 billion in public funds in nanotechnology; private investment was slightly higher ($8.6 billion) (Shapira & Wang, 2010). Nanotechnology is highly globalized in terms of research and development, technology transfer, product engineering and application, and manufacture. The United States is the world leader in this regard; the NNI, launched in the closing days of the Clinton administration, grew from approximately $464 million in its initial year of funding (2001) to a budget request for $1.6 billion in 2010 (Roco, 2001, 2009), representing one of the largest government investments in technology since the Apollo program (McCray, 2009; McNeil, Lowe, Mastroianni, Cronin, & Ferk, 2007). One study (Shapira & Wang, 2010) of international collaboration on 61,300 grant-supported nanotechnology publications found that nearly a quarter (23 percent) had co-authors in different countries—although publication was still concentrated in a small number of places.[29]

The list of promised benefits is seemingly endless, with nano "skeptics" classifying much as hype (Berube, 2006). To take but a few often-mentioned examples (Lane & Kalil, 2005; NNI, 2006):

- Low-cost hybrid solar cells that combine inorganic nanorods with conducting polymers, providing a new, low-cost source of energy;
- Targeted drug delivery, achieved by constructing nanoscale particles that migrate and bond with specific types of cancer cells, which are then be selectively destroyed, thereby offering a non-invasive cure for cancer without the toxic side effects of radiation and chemotherapy;
- "Lab-on-a-chip," providing instant diagnosis of multiple diseases in remote field settings, greatly contributing to public health in poor countries where medical facilities are lacking;
- Ultra high-speed computing, thanks to data storage devices based on nanoscale electronics that provide data densities over 100 times that of today's highest density commercial devices;
- Nanoscale filtration with high efficiencies at low costs, providing a solution for air pollution and water contamination;
- Nano-electro-mechanical sensors that are capable of detecting and identifying a single molecule of a chemical warfare agent; and
- Nanocomposite energetic materials that create propellants and explosives with more than twice the energy output of typical high explosives.

Nanotechnology is said to hold great promise for not only developed economies, but for emerging economies as well. One early study (Singer, Salamanca-Buentello, & Daar, 2005, Table 1; see also Salamanca-Buentello et al., 2005) consulted a panel of 63 experts—60 of whom were from low- and middle-income countries—to rank the 10 nanotechnology applications they felt would be of greatest benefit to developing countries over the next decade. In order of ranking (from top to bottom), these were energy storage, production, and conversion; agricultural productivity enhancement; water treatment and remediation; disease diagnosis and screening; drug delivery systems; food processing and storage; air pollution and remediation; construction; health monitoring; and vector and pest detection. Given such promise, it is not surprising to learn that

> one can conclude that nano has the potential to become the flagship of the industrial production methods of the new millennium in developed as well as in the developing world . . . In view of its pervasiveness, it is likely that the magnitude of this new technology at the frontiers of discovery will exceed those of precedent technologies because the intensity of the impact of a phenomenon is positively correlated to its pervasiveness. These, up to now known circumstances suggest that the possible impacts of nanotechnology will even go beyond those of the first Industrial Revolution. (Bürgi & Pradeep, 2006, p. 648)

Yet despite these grandiose claims, nanotechnology has yet to realize its potential, either as a source of innovative products or as an engine of development. As a recent report conducted for the U.S. Department of Commerce Technology Administration concluded, "[I]t is apparent from roundtables, focus groups, and personal interviews with nanotechnology scientists, venture capitalists, businesses, and consultants, there are no 'home runs' in US nanotechnology commercialization at this time" (McNeil et al., 2007, p. 10). In the United States, with some small exceptions,[30] federal funding under the NNI is directed at basic research, with the expectation that any resulting commercial applications will be market-driven.

One reason that nanotechnology has yet to realize its commercial potential lies in funding that favors basic research over commercialization (McNeil et al., 2007). In China, where governments at all levels play somewhat more significant roles in funding for commercial payoff, it is reasonable to ask the question, is China poised to hit a nanotechnology home run?

China's approach to science and technology development—and to nanotechnology in particular, with its emphasis on indigenous innovation—is highly state-centered. Public investment originates at all levels of government, ranging from support for basic research to funding intended for infrastructure acquisition and to promote commercialization. The National Steering Committee for Nanoscience and Nanotechnology was established in 2000 under the direction of the Minister of Science and Technology, and its National Nanotechnology Development Strategy 2001–2010 (similar to, and influenced by, the U.S. NNI) was adopted the following year. Different levels of government play differing roles as well. As one moves from central to provincial to local levels of government funding, the time horizon for return on investment becomes shorter, and there is a tendency to move from intangible (basic research) to tangible (commercial products) results. At the local level especially, government officials expect a quick turnaround in terms of technological development and market applications (Cheng, 2007). Provincial governments are important not only in provinces containing the major cities (such as Beijing and Shanghai), but also in provinces such as Zhejiang, which neighbors Shanghai, that hope to promote their regional universities as major players by setting up collaborative university-industry science centers. Both provincial and local governments can partner with foreign investors, as with the China-Singapore Suzhou Industrial Park Development Corporation. At the local level, various forms of incubation play a key role.[31]

China's state-run firms—which still account for an estimated 43 percent of GDP, despite China's commitment to privatization (OECD, 2005)[32]—tend to be bureaucratic and conservative, shunning potentially risky investments in favor of short-term, more predictable returns. The emerging private sector, including many small and medium enterprises, remains small, under-capitalized, and generally risk-averse. This poses a challenge for the Chinese government's heightened emphasis on leapfrogging development through nanotechnology or other advanced technologies, for which major payoffs may be years in the future. Yet throughout a number of interviews we have conducted with both academics and entrepreneurs, the most pervasive theme to emerge was that of the importance of government funding and support, not only for basic research but also well into commercialization.

The role of private firms and individuals remains limited. Although more than a thousand R&D centers have been established in China by foreign firms, few are engaged in basic research, and almost none in nanotechnology. International collaborations are more promising. These include institutional partnerships between universities and corporations, study-abroad programs (especially postgraduate degrees earned by Chinese in the United States, Japan, and Europe), and efforts to capitalize on Chinese national pride and identification by recruiting overseas Chinese scientists and engineers to return to China. Informal personal ties are also important, as when American professors mentor former graduate students after they return to China. Universities are an especially important component of China's nanotechnology initiative, which, despite government support across the value chain, remains first and foremost research (rather than development) based.

Despite these limitations, China's accomplishments in nanotechnology have been substantial, although almost entirely at the R&D end of the continuum. In terms of nano-related publications, for example, China now rivals the United States in sheer numbers (Appelbaum & Parker, 2008; Lenoir & Herron, 2009), having averaged an annual growth rate of 83 percent between 1998 and 2007 (Huang & Wu, 2010). In terms of quality, China still lags behind; however, the gap is narrowing, at least as measured by the percentage of total publications that are frequently cited (Shapira & Wang, 2009). Even though first-rate research is being done in a handful of labs in China's elite universities, most research is hampered by the previously mentioned challenges that stifle innovation and original thinking. As Shapira and Wang (2009) have concluded, "while Chinese nanotechnology research has scale and increasing quality, the pathways from laboratory research to successful commercialization remain problematic."

In parallel with the rise in quality of research, there is also an increased focus on commercialization in recent years as industry has become aware of its potential in the past ten years. Larger government programs are aimed at applied projects such as nanoscale electronics, solar cells, and materials for energy storage. There have also been some commercialized products for disease diagnosis, such as the use of gold nanoparticles to detect the presence of Hepatitis B with a saliva or blood sample. As a result of such developments private investors are increasingly becoming more interested in nanotechnology (J. Chen, personal communication, May 19, 2010)—at least for near-term end products, as opposed to basic nanomaterials (Y. Zhang, personal communication, May 13, 2010).

Much of the nano-related research done globally occurs in university laboratories, where research tends to proceed slowly. Companies need to see the benefits more quickly, which discourages active university-business partnerships (Y. Zhang, personal communication, May 13, 2010). Moreover, some universities (at least in Beijing, we were told) require a significant portion of any profits that are realized to be reincorporated into the lab's research stream by professors. This policy serves as a disincentive to patent or license research findings and discoveries (J. Chen, personal communication, May 19, 2010; Y. Zhang, personal communication, May 13 2010). Still, we were told that every year Shanghai University applies for between 600 and 800 invention patents, of which perhaps 200 are granted by SIPO; only a small number actually get commercialized (L. Shi, personal communication, May 14, 2010).

Shanghai University also plays an active role in seeking to raise awareness of nanotechnology, distributing research materials, booklets, and pamphlets, as well as a book (*Secrets of the Nano World*) aimed at middle school students. Since 2009 the university has provided a place where companies can get information about the latest research (a 2010 open house, featuring key projects, drew potential investors, although it was related mainly to basic nanomaterials; L. Shi, personal communication, May 14, 2010; S. Yuan, personal communication, May 14, 2010). Shanghai University and Shanghai Normal University both provide opportunities for middle school students to work in their laboratories, so that the students can develop a taste for university research (S. Yuan, personal communication, May 14, 2010).

There are two nanotech-related service platforms in Shanghai. The East China University of Science and Technology has a Pilot Center for Novel Functional Materials, and Shanghai University has a Pilot Center for Nano Functional Materials. These platforms integrate their resources and share them with enterprises that otherwise would lack specific technological capabilities, testing facilities, or talent. Yet despite such platforms, commercialization of nanotechnology remains elusive. Part of the problem lies in the need for a larger number of service platforms and trained professionals to staff them. More fundamentally, although the service platforms can readily provide assistance when it comes to basic nano materials, there remains a gap between research and commercialization in such areas as electronics. University research tends to be driven by the interests of scientists and engineers, rather than commercial needs; startup companies have little capacity to absorb new technologies; venture capital still plays a limited role (L. Shi, personal communication, May 14, 2010; S. Yuan, personal communication, May 14, 2010).

The city of Suzhou, roughly 50 miles west of Shanghai (and less than a half hour by bullet train), boasts a sprawling industrial park (its jurisdiction covers nearly 300 square kilometers, or more than 100 square miles) devoted to high-tech development. Suzhou Industrial Park has the equivalent of a science and technology bureau, as it gains in industry experience. It initially invited Fortune 500 companies to set up factories in Suzhou, providing initial resources; its current approach involves seeding innovation through the inclusion of universities and research centers. It is still stronger on scientific research than commercialization, with fewer marketers, businessman, and other professional service people (L. Chen, personal communication, January 27, 2011).

BioBay, an 86 hectare (213 acre) sector of the park that opened in June 2007, is devoted to innovation in biopharmaceuticals and nanotechnology. Its complex of modern buildings houses some 150 enterprises engaged in R&D, most started by (formerly) overseas Chinese. The 25 nanotech companies focus on such diverse areas as nano-bio (Suzhou Natong BioMed Co, Suzhou Renesis Therapeutics Inc), nano-materials (Suzhou HIwytech Power Co, Suzhou Nano-micro Biotech Co), nano-devices (Suzhou MEM-Sensing Microsystems Co), nano-photonics (Suzhou Nanowin Science and Technology, OptoTrace Technologies Inc), and nano-environmental (Quantum Water Inc., Bionano Technology Co). In the area of biopharmaceuticals, BioBay covers the complete biomedical value chain from the initial discovery of the disease to diagnosis and treatment; its industry clusters provide services such as new drug development, diagnostic technology, medical equipment, R&D outsourcing, and application of nanotechnology (J. Zhang, personal communication, January 2011).

BioBay provides a variety of incentives to attract its most promising tenants, including startup funds (up to RMB 800,000, or $117,000), rent subsidies, and access to nano-characterization and -fabrication facilities. It serves as a sort of incubator for its tenants, holding investor forums and assisting them in applying for city (Suzhou), Provincial (Jiangsu), and national (MOST) innovation funds. Because Suzhou is a major center of high-tech manufacturing, there are also many opportunities for networking and eventual partnerships, fostered by the creation of the Jiangsu Nano Industry Alliance in January 2010 (NanoGlobe Ptd Ltd, 2010). The government of the city of Suzhou has invested significantly in BioBay, providing roughly four fifths of all funding (BioBay, 2010); Suzhou sees itself as the Silicon Valley of China (H. Wang, personal communication, May 11, 2010).

Whether or not China will be able to convert its growing nanotechnology R&D into commercially viable products remains to be seen. China recently convened its new National Guidance and Coordination Committee on Nanoscience and Technology; its head, Bai Chunli, confirmed that the lack of early involvement of enterprises and low efficiency of tech transfer from research to commercialization are among the problems hampering the further development of nanotech in China. As a result, Chinese returnees have become important in commercialization efforts, heading up many small startup firms. Many (if not most) firms, however, are at the low end of the value chain, producing carbon nanotubes for incorporation in higher-end products elsewhere.

Of course the United States and Europe are not that much closer to commercial success at this stage either. Comparing Korea's experience—in which ongoing (and heavy) public investment in research during the 1960s and 1970s contributed centrally to Korea's industrial success today—Huang and Wu (2010) conclude that China may be similarly poised, provided China "further strengthen[s] the industry-academy collaboration in nanotechnology R&D to boost commercialization and application" (p. 22).

CHINA: A TECHNO-NATIONALIST ECONOMIC SUPERPOWER?

China-watcher Minxin Pei (2009), adjunct senior associate in the Asia Program at the Carnegie Endowment, succinctly summarizes (but also rejects; see Chapter 7, this volume) the conventional wisdom on China's rising economic and political power:

> [T]he Middle Kingdom has all the requisite elements of power—an extensive industrial base, a strong state, a nuclear-armed military, a continental-sized territory, a permanent seat on the United Nations Security Council and a large population base—to be considered as Uncle Sam's most eligible and logical equal. Indeed, the perception that China has already become the world's second superpower has grown so strong that some in the West have proposed a G2—the United States and China—as a new partnership to address the world's most pressing problems.

A recent article in *Forbes* proclaims that "China has Fully Arrived as a Superpower" (Rein, 2009). According to the author,[33] China has finally emerged "as a hotbed of innovation . . . spending $ 9 billion a month on clean energy research . . . within five years it will become the world's largest producer of solar and wind energy." Evidence for this claim includes the return of Chinese expats who have studied and worked overseas, China's recent acquisition of foreign companies such as Volvo and Hummer, and China's rising global influence; China is now Japan's and Brazil's largest trading partner and conducts extensive trade with the Middle East and Africa, where it is sending workers to build highways, provide infrastructure, and open factories. China's economic clout accompanied by an increased willingness to assert political power, as seen in China's significant role at the Copenhagen climate summit, or the fact that the G-20 (in which China is prominent) is replacing the G-8 as the world's primary economic forum.

Nobel Prize–winning economist Robert Fogel (2010) even predicts that the Chinese economy will reach US$120 trillion by 2040, accounting for 40 percent of the world's GDP, with a per capita income of $85,000—more than double that predicted for the EU. Fogel bases his reasoning in part on China's investment in education, with university enrollment (and the number of Chinese studying abroad) increasing more than 150 percent in the four years following then-President Jiang Zemin's 1998 call for increased enrollments in higher education. China now has some 25 million students enrolled in 1,700 higher education institutions, a fivefold increase from a decade earlier (Adams, 2010).[34] Fogel projects that continued growth in enrollments—albeit at slower rates—will add 6 percent to China's annual growth rate. He also points out that in recent years labor productivity has increased 6 percent annually in industry, services,[35] and even agriculture—a trend that he predicts will continue. All of these are said to result in future growth in consumption—an area where China has been lagging—as China's rapidly growing middle class (now numbering in the hundreds of millions) begin to replace government investment as a key driver of economic growth. A McKinsey Global Institute (2006) report concluded "China's economy is on the verge of an important transition in which its consumers will begin to take their place on the world stage" (p. 9). The study's econometric projections (self-described as "robust")[36] predict that

> over the next two decades, the Chinese economy will gradually begin to move away from its historical investment-led growth model, and China's consumers will begin to play a far greater role in their economy's growth . . . and, between 2006 and 2015, a massive middle class will emerge. This rising middle class will be largely an urban phenomenon, which we project will spread beyond China's large wealthy coastal cities, to smaller cities further inland, thus significantly changing the geography of China's consumer market. . ..As the incomes of China's new middle class rise dramatically, so too will their consumption, making China the third-largest consumer market [behind the United States and Japan] in the world by 2025. (p. 10)

Long-term forecasts, whether based on sophisticated econometric modeling or the hunches of Nobel prize–winning economists, are notoriously suspect, and Fogel's projections assume a degree of *ceteris paribus* that seems excessive even by economists' standards.[37] Despite China's considerable investments in education and basic research, the country also confronts a large number of challenges. China's over-reliance on investment at the expense of consumption has long been seen as a drag on future economic growth (Lardy, 2006). The otherwise bullish McKinsey report (2006), cited earlier, nonetheless noted that as a percentage of GDP, between 1995 and 2005 consumption shrank from 47 percent to 37 percent.[38] China's one-party state can stifle the very innovation that party leaders have made central to its economic planning, leading at least one China-watcher to predict a "coming collapse" (Chang, 2010).[39] As previously mentioned, China's universities and laboratories continue to suffer from a hierarchical structure that stifles innovation and creativity—although this may change as growing number of Chinese scientists and engineers return home after studying and working in the Europe, United States, and Japan. Misconduct in science is another problem: Given the enormous pressures to publish in scientific journals (publications in key journals such as *Science* and *Nature* are often rewarded financially), quantity often trumps quality, and plagiarism and other forms of fraud are reportedly widespread (Cao, 2010; Kao, 2010).[40]

There are additional challenges to China's continued rapid growth (Pei, 2009; see also Greentech, 2009).[41] China remains a predominantly agricultural society, with urbanization growing at only 1 percent a year (and with government policies seeking to discourage the enormous eastward migrations that have fueled both the growth of enormous urban areas and their associated industrial development). Per capita incomes remain a tenth that of the United States or Japan, with large numbers of people lacking access to safe drinking water, healthcare, and adequate education. Despite China's large investments in infrastructure, education, and science and technology, the benefits remain inequitably distributed. The three municipalities with the highest proportion of adults with college or advanced degrees are all coastal cities directly under the central government (Beijing, Tianjin, and Shanghai)—places with access to the best universities and high-tech industries (Qian, 2010). Both domestic and international nanotechnology research efforts (discussed subsequently) are similarly concentrated in eastern China (Li & Shapira, 2011). As China's population ages (17 percent of its population will be over 60 by 2020), demands for healthcare and pensions will impact savings. Moreover, as the world's largest exporter, China is increasingly encountering protectionist resistance—although, as noted already, it is responding by moving its production offshore and concentrating on higher value-added "innovative" development strategies. China's reliance on cheap energy (based on high-polluting old coal technology) has resulted in significant environmental damage, as well as looming shortages of potable water and considerable land degradation.[42] Unemployment has grown; one government report estimates that as some 20 million lost their jobs during the 2008 economic downturn, and as many as 40 percent of China's 6 million college graduates may not find work—a problem that is especially daunting for graduates from lower-tier provincial schools (Ernst, 2009; Jacobs, 2010). Chinese science and technology continues to suffer from a well-known innovation deficit and remains highly dependent on foreign technology. The ability of the government to balance market- and state-led approaches remains unclear, particularly because Chinese firms lack innovative capacity. State-run firms—which still account for an estimated two fifths of GDP, despite China's commitment to privatization—tend to be bureaucratic and conservative, shunning potentially risky investments in favor of short-term, more predictable returns. The emerging private sector, including many small and medium enterprises, remains small, under-capitalized, and generally risk-averse, whereas researchers in university laboratories and CAS Institutes lack business know-how and experience in commercializing their ideas.

In the view of Richard P. Suttmeier (2008), who has long studied China's emergence as a technology leader, "China's technology policy has to navigate between the appeals of participating in global production networks and building up *national* technological capabilities [resulting in a] tension between what might be called techno-nationalism and techno-globalism" (p. 10). Suttmeier suggests that in the highly globalized world of science and technology today, the very notion of a "science superpower" needs to be recast: instead of emerging as a national superpower, China may well emerge as a "super-node": In interesting ways—through its open door policies, its capacity for policy learning from foreign experience, its strategies for international scientific and technological cooperation, and especially through its "scientific diaspora"—China is especially well equipped to become a leading presence, if not a "supernode," in twenty-first-century global networks of research and innovation.

WHAT WILL CHINA'S FUTURE HOLD?

There are other limits on China's ability to emerge as the world's dominant geopolitical power, even if it succeeds in overcoming the previously mentioned limitations to continued advances in its scientific and technological capabilities, and manages to translate these capabilities into sustained economic growth. China confronts what Minxin Pei (2009) terms "geopolitical counter-balancing"—unlike the United States, it is surrounded by a tier of strong regional rival powers (Japan, India, Russia), as well as less—but rising—neighbors (Indonesia, Vietnam, South Korea). In Pei's (2009) view, therefore, at least "for the foreseeable future, China will be, at best, only an economic superpower by virtue of its role as one of the world's greatest trading powers . . . Its geopolitical and military influence, meanwhile, will remain constrained by internal fragilities and external rivalry." Nonetheless, it is useful to speculate on how China's rise might play itself out, if current trends continue into the next decade or so.

We have argued that China's future economic growth will increasingly result from its ability to compete at the top of the value chain, competing on technology as well as cost. This will mean that Chinese brands will prove competitive, both globally, and to China's rapidly expanding middle class (now estimated in the hundreds of millions). Chinese products will increasingly be designed, marketed, manufactured, and sold by Chinese firms to Chinese consumers, rather than by foreign multinationals. In short, China may well be poised to make the classic emerging economy transition from export-oriented industrialization to import-substitution industrialization. Such an economic future would have significant foreign policy implications.[43]

At present, China and its trading partners—most notably the United States—are highly interdependent. The U.S. trade deficit with China accounts for a significant portion of China's foreign account surplus, a *quid pro quo* that despite occasional protests on both sides has proven extremely beneficial for both parties. On the Chinese side, the surplus has helped to finance the massive investments in science, technology, and infrastructure that underlie China's rapid economic growth. On the U.S. side, it has kept inflation low by providing consumers with an endless stream of low-cost goods.[44] China's holdings of U.S. Treasury securities—approaching $1.2 trillion as of May 2011, and accounting for nearly a quarter of all foreign Treasury holdings (U.S. Federal Reserve Board, 2011)—gives China a vested interest in the continued stability of the dollar, and the United States an interest in China's economic stability, it has emerged as a major financier of the growing U.S. federal deficit: Both are, to use a well-worn phrase, too big to fail.[45] But are they? As China becomes less dependent on foreign technology, foreign multinationals as a source of employment, and foreign consumers for Chinese-made products, its leaders will no longer see their country's fate as tightly coupled with that of the United States as they have in the past. Foreign markets need no longer be the principal engine of economic growth; foreign currencies no longer the principal source of funding. Although China cannot uncouple its economy from that of the United States precipitously without jeopardizing the value of its dollar reserves, in the long run, such an uncoupling may well occur, as China moves up the value chain—increasingly designing and marketing its own high-technology products, selling to its own growing internal market and off-shoring its low-cost, low-wage manufacturing to Vietnam and other impoverished countries in Africa and elsewhere.

In other words, to the extent that present trends continue, we may expect China to increasingly act as a great power, using its economic and political influence to attempt to shape world events in its own interest. How this would play out is beyond the scope of this paper, but we can speculate on two vastly different possible scenarios.[46]

In the *globalization scenario*, China's global economic interdependence continues to grow, with China becoming less economically dependent on the United States. China increasingly diversifies its trade with other countries, especially the emerging economies in Asia, Latin America, and Africa. Under this scenario, China's dollar reserves are increasingly replaced by Euros, with China pressing for the RMB to become a global currency reserve. China's uncoupling from the United States is managed slowly, so as to avoid jeopardizing the value of its U.S. currency reserves, but over time China has greatly diversified its financial portfolio. The world becomes a network of competing yet interlinked major economic powers, including the United States, Europe, China, Japan, India, Brazil, and perhaps Russia. Under the aegis of the World Trade Organization, China continues on its path of economic liberalization, accepting the core rules and norms of the international system. Under the most optimistic version of this scenario, China follows the path of South Korea, with political democracy following economic liberalization. Building on Ohmae's (1985, 1999) notion of "triad power" in which Europe, Japan, and the

United States determine the shape of the global economy, Glazel, Debackere, and Meyer (2008) suggest that perhaps China should be considered a fourth node vying for dominance in the knowledge economy, along with the United States, Europe, and Japan. Their findings suggest that China is beginning to emerge as a leader in terms of its research profile through heightened publication and citation standards, as we have described above, ultimately transforming the old "triad" in to a "tetrad," (Parker et al., 2009).

In the *geopolitical scenario*, China increasingly flexes its economic and political muscle. With its large population and territory, and a Communist Party elite determined to maintain its power through a combination of economic growth and authoritarian rule, China increasingly decides to "go it alone," relying on its growing internal market, natural resources, and technological capability. In this scenario, China uncouples its economy from that of the United States, increasing trade (including its own exports) with other countries. China comes to rely increasingly on asymmetrical relations with emerging economies, in which it serves as a kind of hegemonic power, providing infrastructure and investment while extracting wealth from cheap labor as well as natural resources. To enhance its growing global power, China develops its military: its navy and air force to protect trade routes and secure itself against real (or imagined) depredation from regional rivals; the People's Liberation Army to secure Chinese interests abroad while assuring quiescence at home. Geopolitical rivalries reassert themselves, and—in the most pessimistic scenario—a wave of protectionism once again threatens global stability.

Will the "Middle Kingdom," after conceding technological, economic, and political leadership to Western powers for some five centuries, reassert what it has come to regard as its historic place among nations: first among equals? Or will China work with other countries to help sustain global economic growth, addressing such common challenges as global climate change, and rest content to play a leading role in the G-20, focusing on its own internal economic and environmental challenges rather than seeking to establish itself as an economic, political, and military force?

At least part of the answer will rest with the United States and other leading powers. If the United States engages in China-bashing in response to its own economic challenges, it seems likely that China will respond both defensively and assertively. China's advances in science and technology can be seen as a threat to U.S. dominance—or as an opportunity for a global science and engineering effort to solve common problems. China's growth trajectory seems clear; how it plays out geopolitically will depend in large part on the response of others.

ACKNOWLEDGMENTS

This material is based upon work supported by the National Science Foundation under Cooperative Agreement No. SES 0531184. Any opinions, findings, and conclusions or recommendations expressed in this material are those of the authors and do not necessarily reflect the views of the National Science Foundation. It was conducted under the auspices of the UCSB Center for Nanotechnology in Society (www.cns.ucsb.edu).

NOTES

1. The argument that a new Beijing Consensus is replacing the Washington Consensus was forcefully made in Ramos (2004), who coined the term.
2. The 863 Program (Key Technologies Research and development Program) provides grants on a competitive basis for applied research in designated sectors; the 973 Program (National Basic Research Program) funds basic research.
3. We have also been told that most patents "sleep in the safe," having been taken out to satisfy government funding sources or enable China to demand licensing fees or thwart potential foreign patents (e.g., see Wadwha, 2011).
4. A special issue of *Nature Magazine* (fall 2004) featured leading Chinese scientists, including some outside of China, arguing against letting government officials determine how research funding should be spent. There was strong concern that the Ministry of Science and Technology (MOST) would have too much power, with some suggesting that MOST be abolished altogether (see McGregor, 2010).
5. This theme was taken from the then CCP General Secretary Jiang Zemin's report to the 15th CCP Congress in 1997, which reads, "We should formulate a long-term plan for the development of science from the needs of long-range development of the country, taking a panoramic view of the situation, emphasizing key points, *doing what we need and attempting nothing where we do not*, strengthening fundamental research, and accelerating the transformation of achievements from high-tech research into industrialization" (emphasis added). This was in turn adapted from the May 1995 decision of the CCP and the State Council to push forward China's S&T progress, although the wording was slight different—"catching up what we need and attempting nothing where we do not" (*you suo gan, you suo bu gan*).
6. The other *science megaprojects* are development and reproductive biology, protein science, and quantum research. It seems questionable whether or not nanotechnology is currently regarded as a science megaproject, because much of the focus and funding is on engineering applications (Cao, 2012).
7. The engineering megaprojects are advanced numeric-controlled machinery and basic manufacturing technology; control and treatment of AIDS, hepatitis, and other major diseases; core electronic components, high-end generic chips, and basic software; drug innovation and development; extra large scale integrated circuit manufacturing and technique; genetically-modified new organism variety breeding; high-definition earth observation systems; large advanced nuclear reactors; large aircraft; large-scale oil and gas exploration; manned aerospace and moon exploration; new-generation broadband and mobile telecommunications; and water pollution and control and treatment.
8. The frontier technology programs are advanced energy; advanced manufacturing; aerospace and aeronautics; biotechnology; information; laser; new materials; and ocean.
9. The key areas are agriculture, energy, environment, information technology and modern services, manufacturing, national defense, population and health, public securities, transportation, urbanization and urban development, and water and mineral resources.

10. Other areas will indirectly fuel high-tech innovation and development—for example, health, culture, and education; and ecological environment.
11. By some accounts China's stimulus spending paid off, at least in terms of jobs created, the real estate market, and overall economic growth (which reportedly reached an annualized rate of nearly 15 percent in the second quarter of 2009) (Bradsher, 2009).
12. The Chinese character for returnees (*hai gui*) sounds like the character for sea turtles, who return to their birthplaces to lay their eggs.
13. One "Thousand Talents" 2009 call for recruitment at the Shanghai Advanced Research Institute required that applicants have an overseas Ph.D., be less than 55 years old, hold a professorship or equivalent position "at famous overseas universities or research institutions," and agree to work in China for a minimum of 6 months each year. The benefits were considerable: startup research fund of no less than 2,000,000 RMB (roughly $300,000), either a salary with performance bonus or a fixed annual salary of 500,000–800,000RMB ($73,000–$117,000), and a housing subsidy of 1,000,000RMB ($150,000) (CAS, 2009b).
14. Recent prominent returnees include Robin Li, co-founder and chairman of Baidu (China's top search engine) and Min fan, CEO of Ctrip (China's self-proclaimed "Premier Travel Site"), both listed on the NASDAQ. Stories also abound about the return of top-level academics, lured back both by the promise of opportunity, handsome laboratory or administrative packages, and national pride (e.g., see Paddock, 2010; Wines, 2011).
15. As Pinto explained, "We're doing R&D in China because they're becoming a big market whose needs are different from those in the U.S . . . energy will become the biggest business for the company. . .[China] will be the biggest solar market in the world" (Bourzac, 2009).
16. Shi Yigong's Ph.D. in biophysics was earned at John's Hopkins in 1995; he was on the faculty at Princeton University prior to joining Tsinghua University, his undergraduate alma mater, in 2003. Rao Yi, a molecular biologist and life science dean at Peking University, returned to China in 2007 (he at the time held a chair at Northwestern University) after spending nearly a quarter century in the United States. In an interview Rao said he returned because he "didn't want to just look at the epoch making changes" that were occurring in China (Tang, 2010).
17. Officially, the "Circular on Carrying Out the Work on Accreditation of National Indigenous Innovation Products."
18. Indigenous innovation was operationalized as product innovations that had been patented or trademarked, in China, by a Chinese company.
19. China has never signed on to the WTO Agreement on Government Procurement, whose "purpose is to open up as much of this business as possible to international competition. It is designed to make laws, regulations, procedures and practices regarding government procurement more transparent and to ensure they do not protect domestic products or suppliers, or discriminate against foreign products or supplier" (WTO, 2010). Thus far 40 countries have signed the agreement, including the 27 members states of the EU and the United States. (No emerging economy has signed on, with the exception of Korea, Singapore, and Taiwan.)
20. Partly in response to foreign business concerns (some three dozen foreign associations lobbied the Chinese government, along with global Chinese companies with foreign partners), Circular 618 was modified in April 2010, removing the most onerous requirements (Kennedy, 2010; McGregor, 2010). China also announced that it plans to join the WTO Agreement on Government Procurement "as soon as possible"—a commitment that is regarded with some skepticism by its critics (Segal, 2010).

21. Utility patents apply to "any new technical solution relating to the shape, the structure, or their combination, of a product which is fit for practical use." Utility patents are granted largely on the basis of filing proper paperwork, are quickly granted, and provide protection for 10 years (Babel, 2008).
22. In 2009, of nearly 1 million patent applications that were filed with SIPO, 90 percent were from Chinese nationals. Among these, slightly less than a third were utility model patents, whereas slightly more than a third were design model patents (McGregor, 2010).
23. The U.S. government has raised the issue at the Strategic and Economic Dialogue, "the most important meeting between the two countries" (Segal, 2010).
24. The Commission (2010) was created in 2000 with a mandate "to monitor, investigate, and report to Congress on the national security implications of the bilateral trade and economic relationship between the United States and the People's Republic of China" (p. iii).
25. Yet at the same time, a 2010 survey by the American Chamber of Commerce in China found that only 2 percent reported a business negative outlook in China over the next five years—the lowest figure ever reported (Kennedy, 2010).
26. In recent years China has accounted for nearly half of the U.S. trading deficit in goods. The Commission (2010) notes "for the first eight months of 2010, China's goods exports to the United States were $229.2 billion, while US goods exports to China were $55.8 billion, with the US trade deficit in goods at $173.4 billion, an increase of 20.6 percent over the same period in 2009 ($143.8 billion). This constitutes a four-to-one ratio of Chinese exports to its imports from the United States" (p. 2).
27. It is beyond the scope of this paper to analyze China's growing military capabilities. One recent detailed analysis, however, gave mixed reviews: "In conflicts around the country's borders, when China can compensate for technological inferiority by using more traditional warfare methods, such as flooding the frontline with masses of soldiers, the weakness of its procurement process may have a relatively limited impact on its actual military capability. On the other hand, in remote and complex conflicts, when combined technological capabilities—for example, sophisticated air and naval systems, precise guided weapons (PGW) and sophisticated [Command, Control, Communication, Computers, Intelligence, Surveillance, and Reconnaissance] systems—play a decisive role, a problematic procurement process may have a negative impact on China's ability to achieve its military objectives (Evron, 2010).
28. A nanometer is a billionth of a meter, roughly equal to 3–6 atoms side-by-side. A human hair is roughly 80,000 nanometers wide.
29. China had the largest number of publications (14,500), followed by the United States (13,800), Germany and Japan (3,800 each), and the rest of the EU (3,500). Although China is now the world's largest producer of scientifically indexed journal articles in the field of nanotechnology, the quality of its output (as measured by the number of times its articles are cited) is of lower quality. By far the densest network of international collaborations among all countries is between the United States and China (Shapira & Wang, 2010).
30. The U.S. government's Small Business Administration runs two programs that support small businesses: The Small Business Innovation Research Program (SBIR), which requires all federal agencies that finance R&D to allocate 2.5 percent of their budgets for projects involving small businesses, and the Small Business Technology Transfer Program (http://www.sbir.gov/about/index.htm). These two programs allocate some $2 billion annually in

what amounts to federally funded venture capital for high-tech small businesses. For an example of one nanotech startup that has benefited from SBIR funding, see Parker and Appelbaum (2010).

31. For a more detailed discussion of different centers that promote nanotechnology research, development, and commercialization, see Huang and Wu (2010).
32. In 1997 President Jiang Zemin called for privatization (*feigongyou*, or "non-public ownership") of state-owned enterprises, a plan that was ratified by the Ninth National People's Congress the following year.
33. Shaun Rein is Managing Director of CMR, a strategic market intelligence firm headquartered in Shanghai, and a columnist for *Forbes* on leadership, marketing, and China; he also writes for *Business Week*'s "Asia Insight" section. He was previously Chief of Research for Inter-Asia Venture Management, a venture capital firm, and Managing Director (and Country Head director for China) for WebCT.
34. When the Program for International Assessment (PISA) test was administered via OECD to 15-year-old students in 65 countries, a representative sample of 5,100 Shanghai students scored the highest on all three tests (science, reading, and mathematics). U.S. students came in 23rd, 17th, and 31st on the three tests, respectively. While Shanghai is clearly not representative of all China, it is interesting to note that on the math and science tests, all Chinese regions that participated in the test outperformed the United States (Dillon, 2010).
35. Fogel also argues that the growth of the service sector is most likely considerably under-estimated in official statistics, because an unknown part of the Chinese economy is off the books; moreover, he argues that it is difficult to measure improvements in services such as education and health care.
36. For their "China Consumer Demand Model (ver 20.0)," the study drew on macroeconomic, socio-demographic, household income/expenditure, and price data from a variety of Chinese and international sources. Details about the data and model ate provided in Appendix B of the report. See Pettis (2009), however, for a critique of the McKinsey forecast.
37. Fogel acknowledges that China faces demographic challenges resulting from its rapidly aging society, conceding that "sceptics point to many obstacles that could derail the Chinese bullet train over the next 30 years: rising income inequality, potential social unrest, territorial disputes, fuel scarcity, water shortages, environmental pollution, and a still-rickety banking system." He concludes, however, that "although the critics have a point, these concerns are no secret to China's leaders; in recent years, Beijing has proven quite adept in tackling problems it has set out to address."
38. China's 37 percent is extremely low by world standards. Consumption for European economies typically runs 55–65 percent; the United States, 70–72 percent; emerging economies in Latin America, 65–70 percent; Asian economies, 50–60 percent (Pettis, 2009). Based on these figures, some scepticism is in order; as Pettis notes, consumption in China would have to grow faster than GDP to bring Chinese consumption in line with that of other economies; given China's high level of (investment-led) GDP growth, this may prove challenging.
39. Gordon Chang is the author of *The Coning Collapse of China* (Random House, 2001). Since the publication of Chang's book, China's economy has approximately doubled in size.
40. Apart from some well-publicized cases—such as Jinggangshan University scientists who fabricated 70 papers submitted to *Acta Crysytallographica*—a recent survey by China's leading scientific association found that half the

respondents claimed they had colleagues who had engaged in misconduct (Cao, 2010).

41. A recent commentary in the *Beijing Times*, as reported in *People's Daily* (Li, 2011) points to four "uncertainties" about China's future: continued reliance on massive public infrastructure investments, as symbolized by high speed rail (plagued by high costs and low usage); the shaky economic condition of many small and medium-sized enterprises; inflation (estimated at nearly 7 percent in July 2011); and signs of slowing GDP growth in China's major cities.
42. China's efforts to counter its environmental problems are well known. China officially plans to produce a fifth of its energy from renewable sources such as wind and solar by 2020, matching Europe's goal, according to Zhang Ziaoqiang, vice chairman of China's National Development and Reform Commission (NRDC). These goals are to be achieved by investing much of China's stimulus money in low-carbon energy sources, the creation of a more efficient energy transmission system, and carbon-efficient transportation systems (Borger & Watts, 2009).
43. Such a future is likely 10–20 years off. A trio of recent thoughtful books about China's future conclude that China is clearly on the economic and political rise, even as it faces challenges that could alter its trajectory—and that constructive U.S. engagement with China would help assure a peaceful transition to China's increased role. See Ross and Feng (2008), Bergsten, Freeman, Lardy, and Mitchell (2008), and Lampton (2008).
44. The downsides for both parties are also obvious: for China, a vast supply of funding for speculative investments in housing, office space, Olympic venues, and infrastructure that may prove to be a bubble; on the U.S. side, the loss of manufacturing jobs. While American blue collar workers may have suffered as a result, all Americans as consumers have benefited.
45. By summer 2011 China had amassed an estimated $3 trillion in foreign reserves (Bradsher, 2011) to invest in high-tech development initiatives; roughly half of these were invested in U.S. treasuries, contributing to an extraordinarily high degree financial interdependence between the two countries. It is worth noting that Japan's holdings of U.S. Treasury securities are also substantial, totaling $912.4 billion in May 2011. China and Japan together account for nearly half (46 percent) of all foreign holdings (U.S. Federal Reserve Board, 2011).
46. For a more detailed examination of some possible implications of China's economic growth for both the United States and China's regional neighbors, see Appelbaum and Parker (2012).

REFERENCES

Adams, J. (2010, January 6). Get ready for China's domination of science. *New Scientist*. Retrieved from http://www.newscientist.com/article/mg20527426.900-get-ready-for-chinas-domination-of-science.html

Adams, J., King, C., & Ma, N. (2009, November). *Global research report: China*. Thomas Reuters. Retrieved from http://science.thomsonreuters.com/m/pdfs/grr-India-oct09_ag0908174.pdf

Appelbaum, R. P., & Parker, R. A. (2008). China's bid to become a global nanotech leader: Advancing nanotechnology through state-led programs and international collaborations. *Science and Public Policy, 35*(5), 319–334.

Appelbaum, R. P. & Parker, R. A. (2012). China's move to high-tech innovation: Some regional policy implications. In C. Dent & J. Dosch (Eds.), *The Asia-Pacific, regionalism and the global system*. Cheltenham, UK: Edward Elgar.

Appelbaum, R. P., Parker, R. A., & Cao, C. (2011). Developmental state and innovation: Nanotechnology in China. *Global Networks, 11*(3), 298–314.

Appelbaum, R. P., Parker, R. A., Cao, C., & Gereffi, G. (2011). China's (not so hidden) developmental state: Becoming a leading nanotechnology innovator in the 21st century. In F. Block & M. R. Keller (Eds.), *States and innovation: The US government's role in technology development*, pp. 217–235. Boulder, CO: Paradigm Press.

Babel, T. S. (2008). Patents in China—is there any real protection? *WRAL Techwire.* Retrieved February 1, 2011, from http://wraltechwire.com/business/tech_wire/opinion/story/2776264/

Bai, C. L. (2005, April 20). Address from the director. Website for China's National Center for Nanoscience and Technology. Retrieved from http://www.nanoctr.cn/e_view.jsp?tipid=1116405057331

Bai, C. L., & Wang, C. (2007). Nanotechnology research in China. In L. Jakobson (Ed.), *Innovation with Chinese characteristics: High-tech research in China*, pp. 71–98. New York: Palgrave MacMillan.

Batelle. (2009, December). 2010 Global R&D funding forecast. *R&D Magazine.* Retrieved from http://www.rdmag.com/uploadedFiles/RD/Featured_Articles/2009/12/GFF2010_ads_small.pdf

Bergsten, C. F., Freeman, C., Lardy, N. R., & Mitchell, D. J. (2008). *China's rise: Challenges and opportunities.* Washington, DC: Peterson Institute for International Economics.

Berube, D. (2006). *Nano-hype: The truth behind the nanotechnology buzz.* Amherst, NY: Prometheus Books.

BioBay. (2010). Zhao, Jennifer, Deputy Director, Investor Service Division, Economic & Trade Development Bureau, Suzhou Industrial Park Administrative Committee; Jing Sun, International Cooperation Division, Science & Technology Development Bureau, Suzhou Industrial Park Administrative Committee; Mary Zhao, Business Development Specialist, BioBay; Jennifer Wei, Senior Specialist, Industrial Service Department, BioBay (group interview, May 11).

Borger, J., & Watts, J. (2009, June 10). China launches green power revolution to catch up on West. *The Guardian.* Retrieved March 2012, from http://www.guardian.co.uk/world/2009/jun/09/china-green-energy-solar-wind

Bourzac, C. (2009, December 22). Applied Materials moves solar expertise to China. MIT: *Technology Review.* Retrieved March 2012, from http://www.technologyreview.com/article/24274/)

Bradsher, K. (2009, September 18). Recovery picks up in China as U.S. still ails. *New York Times.* Retrieved March 2012, from http://www.nytimes.com/2009/09/18/business/global/18yuan.html?pagewanted=all

Bradsher, K. (2011, August 8). Chinese fault Beijing over foreign reserves. *New York Times.* Retrieved March 2012, from http://www.nytimes.com/2011/08/09/business/global/chinese-fault-beijings-moves-on-foreign-reserves.html?ref=china

Breakthrough Institute. (2009, November). *Rising tigers sleeping giant: Asian nations set to dominate the clean energy race by out-investing the United States.* New York: Author. Retrieved March 2012, from http://thebreakthrough.org/blog/Rising_Tigers.pdf

Bürgi, B. R., & Pradeep, R. (2006, March). Societal implications of nanoscience and nanotechnology in developing countries. *Current Science, 90*(5), 645–658.

Cao, C. (2010, January 18). A climate for misconduct. *New York Times* blog, Room For Debate: Will China Achieve Supremacy? Retrieved March 2012, from http://roomfordebate.blogs.nytimes.com/2010/01/18/will-china-achieve-science-supremacy/?partner=rss&emc=rss

Cao, C. (2012). *Commercialization of nanotechnology in China.* CNS working paper. Santa Barbara, California: Center for Nanotechnology in Society.

Cao, C., Suttmeier, R. P., & Simon, D. F. (2006, December). China's 15-year science and technology plan. *Physics Today,* 38–43.

China Academy of Sciences. (2009a, September 29). "Thousand talents program" draws 204 more overseas science talents. [Press release]. Retrieved March 2012, from http://english.cas.cn/Ne/CASE/200910/t20091014_45162.shtml

China Academy of Sciences. (2009b, November 25). "Thousand talents program" recruitment at Shanghai Advanced Research Institute [Press release]. Retrieved March 2012, from http://english.cas.cn/JU/PO/200911/t20091125_47648.shtml

Chang, G. (2010, January 18). Hard sciences require freedom too. *New York Times* blog, Room for debate: Will China achieve supremacy? Retrieved March 2012, from http://roomfordebate.blogs.nytimes.com/2010/01/18/will-china-achieve-science-supremacy/?partner=rss&emc=rss

Dillon, S. (2010, December 7). Top test scores from Shanghai stun educators. *The New York Times*. Retrieved March 12, from http://www.nytimes.com/2010/12/07/education/07education.html

Ernst, D. (2009). *Asia's challenges in the global crisis—a catalyst for change?* Honolulu, HI: East-West Center. Retrieved March 2012, from http://www.eastwestcenter.org/news-center/east-west-wire/asias-challenges-in-the-global-crisis-a-catalyst-for-change/

Evron, Y. (2010). *China's military build-up in the early twenty-first century: From arms procurement to war-fighting capability*. S. Rajaratnam School of International Studies, Singapore, RSIS Working Paper, 218. Retrieved March 2012, from http://www.rsis.edu.sg/publications/WorkingPapers/WP218.pdf

Fogel, R. (2010, January). $123,000,000,000,000: China's estimated economy by the year 2040. Be warned. *Foreign Policy*. Retrieved March 2012, from http://www.foreignpolicy.com/articles/2010/01/04/123000000000000

Frost & Sullivan recognizes Suntech Power for its technical expertise with low cost, high quality products for solar energy markets. (2008, April 15). *Business Wire*. Retrieved March 2012, from http://www.redorbit.com/news/business/1342480/frost__sullivan_recognizes_suntech_power_for_its_technical_expertise/index.html

Gabriele, A. (2009, April 5). *The role of the state in China's industrial development: A reassessment* (MPRA Publication No., 14551). Retrieved March 2012, from http://ideas.repec.org/p/pra/mprapa/14551.html

Glanzel, W., Debackere, K., & Meyer, M. (2008). "Triad" or "Tetrad"? On global changes in a dynamic world. *Scientometrics, 74*(71), 71–88.

Greentech. (2009). *The China greentech report 2009*. The China Greentech Initiative, MangoStrategy LLC. Retrieved March 2012, from http://www.china-greentech.com/report/

Grueber, M., & Studt, T. (2009, December 22). Emerging economies drive global RT&D growth. *R&D Magazine*. Retrieved March 2012, from http://www.rdmag.com/Featured-Articles/2009/12/Policy-And-Industry-Global-Funding-Report-Emerging-Economies-Drive-Global-R-D-Growth/

Holman, M. J. K., Nordan, M., Sullivan, T., Mamikunian, V., Nagy, C., Lackner, D., Bünger, M., Biegala, T., Jabbawy, B., Yoo, R., Kusari, U., & Dobbins, M. (2007). *The nanotech report* (5th ed.). New York: Lux Research, Inc.

Huang, C., & Wu, Y. (2010). A sure bet or a scientometric mirage? As assessment of Chinese progress in nanotechnology. Unpublished paper, UNU-MERIT and Masticht University.

Huang, Y. (2011, January). Rethinking the Beijing consensus. *Asia Policy, 11*, 1–26. Retrieved March 2012, from http://www.nbr.org/publications/asia_policy/AP11/AP11_B_BeijingConsensus.pdf

Jacobs, A. (2010, December 11). China's army of graduates struggles for jobs. *The New York Times*. Retrieved March 2012, from http://www.nytimes.com/2010/12/12/world/asia/12beijing.html

Kao, J. (2010, January 18). Can quantity lead to quality? *New York Times* blog, Room For Debate: Will China Achieve Supremacy? Retrieved March 2012, from http://roomfordebate.blogs.nytimes.com/2010/01/18/will-china-achieve-science-supremacy/?partner=rss&emc=rss

Kennedy, S. (2010, September). Indigenous innovation: Not as scary as it sounds. *China Economic Quarterly, 15–20.*

LaFraniere, S. (2010, January 7). Fighting trend, China is luring scientists home. *New York Times*. Retrieved March 2012, from http://www.nytimes.com/2010/01/07/world/asia/07scholar.html

Lampton, D. M. (2008). *The three faces of Chinese power: Might, money, and minds*. Berkeley: University of California Press.

Lane, N., & Kalil, T. (2005). The national nanotechnology initiative: Present at the creation. *Issues in Science and Technology*. Retrieved March 2012, from http://www.issues.org/21.4/lane.html

Lardy, N. R. (2006, October). *China: Towards a consumption-driven growth path*. Institute for International Economics Policy Briefs in International Economics (Publication No. PB06–6).

Lenoir, T., & Herron, P. (2009). Tracking the current rise of Chinese pharmaceutical bionanotechnology. *Journal of Biomedical Discovery & Collaboration, 4*, 1–38.

Li, L. (2011, August 11). China must be wary of four uncertainties. *People's Daily*. Retrieved March 2012, from http://english.people.com.cn/90777/7566102.html#

Li, T., & Shapira, P. (2011). Regional development and interregional collaboration in the growth of nanotechnology research in China. *Scientometrics, 86*, 299–315.

Lieberman, J. (2005). Foreword. In L. E. Foster (Ed.), *Nanotechnology: Science, innovation, and opportunity, pp. xi*. Upper Saddle River, NJ: Prentice Hall.

McCray, W. P. (2009). From lab to iPod: A story of discovery and commercialization in the post-cold war era. *Technology and Culture, 50*(1), 58–81.

McGregor, J. (2010). *China's drive for indigenous innovation: A web of industrial policies*. U.S. Chamber of Commerce Global Regulatory Cooperation Project. Retrieved March 2012, from http://www.uschamber.com/sites/default/files/reports/100728chinareport_0.pdf

McKinsey Global Institute. (2006, November). *From "made in China" to "sold in China": The rise of the Chinese urban consumer*. McKinsey Global Institute. Retrieved March 2012, from http://www.mckinsey.com/mgi/publications/china_consumer/index.asp

McNeil, R. D., Lowe, J., Mastroianni, T., Cronin, J., & Ferk, D. (2007, September). *Barriers to nanotechnology commercialization*. Final report prepared for the U.S. Department of Commerce Technology Administration. Retrieved March 2012, from http://www.google.com/url?sa=t&source=web&ct=res&cd=1&ved=0CAcQFjAA&url=http percent3A percent2F percent2Fwww.ntis.gov percent2Fpdfpercent2FReport-BarriersNanotechnologyCommercialization.pdf&ei=deVES43JOZ7KMdPQ-bwF&usg=AFQjCNGll0L9NsmTs2jY9Vu3mGCrE9ESFg&sig2=OZ05h4DpubhW5kZxSTd27A

Ministry of Science and Technology, United Office of High-Tech Program (MOST). (2001). Overview of 863 Program in the tenth five-year plan period. *Annual report 2001: The National High Technology Research and Development Program of China (863 Program)*. Retrieved March 2012, from http://www.863.org.cn/english/annual_report/annual_repor_2001/200210090007.html

Ministry of Science and Technology, United Office of High-Tech Program (MOST) (2005). *National high-tech R&D program (863 Program)*. Retrieved March 2012, from http://bg.chineseembassy.org/eng/dtxw/t202501.htm

Naik, G. (2009, December 21). R&D spending in US expected to rebound. *The Wall Street Journal*. Retrieved March 2012, from http://online.wsj.com/article/SB10001424052748703344704574610350092009062.html

NanoGlobe Ptd Ltd. (2010, February 3). China Suzhou BioBay—young and progressive nano-bio incubator. *Nanotechnology Now*. Retrieved March 2012, from http://www.nanotech-now.com/columns/?article=407

Nanotechnology Industrialization Base of China (NIBC). (2006, August 6). Brochure provided by the Nanotechnology Industrialization Base of China during interview.

OECD. (2005, September). Policy brief: China's governance in transition. *OECD Observer*. Retrieved March 2012, from http://www.oecd.org/dataoecd/49/13/35312075.pdf

Ohmae, K. (1985). *Triad power: The coming shape of global competition*. New York: The Free Press.

Ohmae, K. (1999). *The borderless world: Power and strategy in the interlinked economy*. New York: Harper.

Oster, S., Shirouzu, N., & Glader, P. (2010, December 28). China squeezes foreigners for share of global riches. *The Wall Street Journal: Asia Business*. Retrieved March 2012, from http://online.wsj.com/article/SB10001424052970203731004576045684068308042.html

Paddock, R. C. (2010, September 23). Booming China lures key professors home from US. *AOL News*. Retrieved March 2012, from http://www.aolnews.com/2010/09/23/booming-china-lures-key-professors-home-from-us/

Parker, R., Ridge, C., Cao, C., and Appelbaum, R. (2009). "China's Nanotechnology Patent Landscape: An Analysis of Invention Patents Filed with State Intellectual Property Office." *Nanotechnology Law and Business Review* 6, 524–539.

Pei, M. (2009, December 29). China's not a superpower. *The Diplomat Magazine*. Retrieved March 2012, from http://www.the-diplomat.com/001f1281_r.aspx?artid=357

Pettis, M. (2009, December 5). The difficult arithmetic of Chinese consumption. *China Financial Markets*. Retrieved March 2012, from http://mpettis.com/2009/12/the-difficult-arithmetic-of-chinese-consumption/

Porter, A. L., Newman, N. C., Roessner, J. D., Johnson D. M., & Jin, X. (2009). International high tech competitiveness: Does China rank #1? *Technology Analysis and Strategic Management, 21*(2), 173–193. Retrieved March 2012, from http://www.tpac.gatech.edu/papers/HTI_China1_2008_jun10.pdf

Qian, H. (2010). Talent, creativity and regional economic performance: The case of China. *American Regional Science, 45*, 133–156.

Ramos, J. C. (2004). *The Beijing consensus*. London: Foreign Policy Centre.

Rein, S. (2009, December 19). Yes, China has fully arrived as a superpower. *Forbes*. Retrieved March 2012, from http://www.forbes.com/2009/12/15/china-superpower-status-leadership-citizenship-trends.html

Renn, O., & Roco, M. C. (2006). Nanotechnology and the need for risk governance. *Journal of Nanoparticle Research, 8*, 2.

Roco, M. C. (2001). From vision to the implementation of the US National Nanotechnology Initiative. *Journal of Nanoparticle Research, 3*(1), 5–11.

Roco, M. C. (2009, November 4–6). Global development and governance of nanotechnology. PowerPoint presentation prepared for UCSB-CNS *Emerging Technologies/Emerging Economies Conference*. Washington, DC: Woodrow Wilson International Center for Scholars.

Roco, M. C., Williams, S., & Alivisatos, P. (Eds.). (1999). *Nanotechnology research directions: IWGN workshop report: Vision for nanotechnology research and development in the next decade*. National Science and Technology Council, Committee on Technology, Interagency Working Group on Nanoscience, Engineering and Technology. Loyola College, MD: WTEC. Retrieved December 15, 2011 from http://www.wtec.org/loyola/nano/IWGN.Research.Directions/

Ross, R. S., & Feng, Z. (Eds.) (2008). *China's ascent: Power, security, and the future of international politics*. Ithaca, NY: Cornell University Press.

Salamanca-Buentello, F., Persad, D. L., Court, E. B., Martin, D. K., Daar, A. S., & Singer, P. A. (2005, April). Nanotechnology and the developing world. *PLoS Medicine* 2(4), e97.

Segal, A. (2010, September 28). China's innovation wall: Beijing's push for homegrown technology. *Foreign Affairs*. Retrieved March 2012, from http://www.foreignaffairs.com/articles/66753/adam-segal/chinas-innovation-wall

Shapira, P., & Wang, J. (2009). From lab to market: Strategies and issues in the commercialization of nanotechnology in China. *Asian Business & Management, 8*, 461–489.

Shapira, P., & Wang, J. (2010). Follow the money. *Nature, 468*, 627–628.

Shi, Y., & Rao, Y. (2010). China's research culture. *Science, 329* (5996), 1128.

Singer, P. A., Salamanca-Buentello, F., & Daar, A. S. (2005). Harnessing nanotechnology to improve global equity. *Issues in Science and Technology*. Retrieved from http://www.issues.org/21.4/singer.html

Suttmeier, R. P. (2007, November 10). Engineers rule, OK? *New Scientist*, 71–73.

Suttmeier, R. P. (2008, December 9–10). *The discourse on China as science and technology superpower: Assessing the arguments*. Paper presented at the International Symposium on China As a Science and Technology Superpower. Tokyo: China Research Center of the Japan Science and Technology Center.

Tang, Y. (2010, June 3). The science of life. *Beijing Review*. Retrieved March 2012, from http://www.bjreview.com.cn/quotes/txt/2010–05/31/content_275381.htm

U.S.–China Economic and Security Review Commission. (2010, November). *2010 report to Congress*. Washington, DC: US Government Printing Office. Retrieved March 2012, from http://www.uscc.gov/annual_report/2010/annual_report_full_10.pdf

U.S. Federal Reserve Board. (2011, July 18). *Major foreign holdings of US Treasury Securities*. US Federal Reserve Board, Department of the Treasury. Retrieved March 2012, from http://www.treasury.gov/resource-center/data-chart-center/tic/Documents/mfh.txt

Valigra, L. (2009, August 27). The innovation economy: China aims to spark indigenous innovation. *Science Business*. Retrieved March 2012, from http://bulletin.sciencebusiness.net/sb/login.php?page=/ebulletins/showissue.php3?page=/548/art/14919&ch=1

Wadwha, V. (2010, January 18). Many reasons to return. *New York Times* blog, Room for debate: Will China achieve supremacy? Retrieved March 2012, from http://roomfordebate.blogs.nytimes.com/2010/01/18/will-china-achieve-science-supremacy/?partner=rss&emc=rss

Wadwha, V. (2011, January 10). China could game the US in intellectual property. *Bloomberg Businessweek*. Retrieved March 2012, from http://www.businessweek.com/technology/content/jan2011/tc2011017_509416.htm

Wines, M. (2011, January 28). A US–China odyssey: Building a better mouse map. *The New York Times*. Retrieved from http://www.nytimes.com/2011/01/29/world/asia/29china.html?partner=rss&emc=rss

Woyke, E. (2010, January 11). China edges into the tech spotlight. *Forbes*. Retrieved March 2012, from http://www.forbes.com/2010/01/11/hisense-ces-electronics-technology-personal-china.html

World Trade Organization. (2010). *The plurilateral agreement on government procurement*. Retrieved February 1, 2011, from http://www.wto.org/english/tratop_e/gproc_e/gp_gpa_e.htm

Xiao, G. (2010, January 18). Strengths from the top. *New York Times* blog, Room for Debate: Will China Achieve Supremacy? Retrieved March 2012, from http://roomfordebate.blogs.nytimes.com/2010/01/18/will-china-achieve-science-supremacy/?partner=rss&emc=rss

Zhou, Q., & Yang, A. (2010, August 10). The fabricated high-tech boom. *Caixin* Online. Retrieved from http://english.caing.com/2010-08-10/100168670.html

Part III

Contesting the Field

Knowledge, Power, and Reflexivity in the Construction of Nanotechnology

9 Nanotechnologies and Upstream Public Engagement

Dilemmas, Debates and Prospects?

Adam Corner and Nick Pidgeon

Nanotechnologies are notoriously difficult to pin down. At the most fundamental level, nanoscientists and regulators struggle with their very definition. Are they technologies constrained within nano dimensions or technologies that contain nanoscale components? Many argue that functionality, rather than size per se, is the defining feature of nanotechnologies—the novel properties that some nano-particles possess (Wood, Jones, & Geldart, 2007).

Despite these unanswered questions, speculation has continued to grow regarding the economic and social impact of nanotechnologies. Nanotechnology advocates promote their potential to revolutionize environmental monitoring (e.g., high quality water purification), energy efficiency (e.g., the "lightweighting" of materials), and the provision of medical treatments (see Oakdene, 2007; Renn & Roco, 2006). For example, Roco (2004) has claimed, "It is conceivable that by 2015, our ability to detect and treat tumors in their first year of occurrence might totally eliminate suffering and death from cancer" (p. 7).[1]

Grand claims like these have been a part of the conversation about nanotechnologies from the outset. Feynman's prescient lecture "There's Plenty of Room at the Bottom" (1959) foreshadowed a world where molecular-level control would produce significant advances in science and engineering, whereas Eric Drexler's (1986) now largely discredited vision of molecular self-assembly fired the minds of technologists and science fiction writers alike. Although some of the initial claims about nanotechnologies (e.g., molecular self-assembly—Drexler, 1986) have faded over time, enthusiasm for the world-changing potential of nanotechnologies has not diminished. A great deal of financial investment is being allocated by national research councils and commercial organizations alike. The market in nanofoods, for example, was projected to reach over US$20 billion by 2010 (Allianz & OECD, 2005). Similarly, the number of nanotechnology patent applications has increased dramatically in the past few years (Nightingale, Smith, van Zwanenberg, Rafols, & Morgan, 2008).

But against this backdrop of optimism about the technological and economic impact of nanotechnologies, a counter-view of cautious uncertainty and in some cases outright opposition has emerged (e.g., Miller & Senjen, 2008). One problem is that the revolutionary potential of nanotechnologies has not yet been realized. Although there are over 1,000 products containing nanotechnology now commercially available, they have tended to be focused on consumer convenience (e.g., translucent sunscreen). Thus, there is a frequently noted disconnect between the rhetoric and reality of nanotechnologies (Wood et al., 2007).

Another problem is that any novel benefits that nanotechnologies might bring must be weighed against the novel risks they may also entail. Currently, there are few signs that nanotechnologies pose a public health or environmental threat (although this has not precluded discussion of precautionary/responsible regulatory action for a range of nanotechnology applications). But one of the reasons that there is little evidence of danger is that there is very little evidence at all, with considerable uncertainty around basic questions of toxicology and exposure (Roco, Mirkin, & Hersam, 2010; Royal Commission on Environmental Pollution, 2008; Royal Society & Royal Academy of Engineering, 2004; Werlin et al., 2011). Beyond these questions of scientific prediction and risk assessment, nanotechnologies raise wider social, ethical, and governance issues. These include concerns over long-term unintended consequences, the means by which governments and society might control the technologies, social risks from covert surveillance arising from nano-based sensors and systems, and financial or other detrimental impacts on the economies of developing and developed countries (Pidgeon & Rogers-Hayden, 2007).

Definitional disputes and competing claims about the risks and benefits of nanotechnologies mean that pursuing public engagement is a major challenge. Currently, the vast majority of people know very little about nanotechnologies: They continue to be "upstream" of significant public awareness. But the implications of nanotechnologies for science and society are potentially far-reaching, and so "upstream engagement" has become a central part of the nanotechnology debate (e.g., see Royal Society & Royal Academy of Engineering, 2004).

Upstream engagement is a term that has come to be firmly embedded in the vocabulary of social researchers and policymakers, and the basic idea is straightforward: For emerging and potentially controversial areas of science, attempts should be to involve members of the public in a dialogue at the earliest possible stage. The preceding discussion of uncertainties over the putative risks and benefits of nanotechnologies should make clear why politicians, concern groups, and industry stakeholders are keen to pursue upstream public engagement on this topic. There is a great deal at stake—not just in terms of economic and technological progress, but also the ethical and social questions that nanotechnologies raise. The potential applications of nanoscience have far-reaching consequences, which means that methodologies for allowing members of the public to engage with (and help shape) these consequences must also be fit for purpose.

Upstream engagement is typically defined as public engagement that occurs before significant research and development of a new technology has begun, public controversy about the topic is not currently present, and entrenched attitudes or social representations have not yet been established (Pidgeon & Rogers-Hayden, 2007). The purpose of this chapter is to review and evaluate existing upstream engagement work on nanotechnologies. Although there is widespread agreement that it is beneficial for experts and policymakers to involve citizens in discussion about emerging areas of science and technology at the earliest possible stage (Hagendijk & Irwin, 2006; Royal Society & Royal Academy of Engineering, 2004), the notion of upstream engagement is not without its critics (Flynn, Bellaby & Ricci, 2009). Even proponents of upstream engagement acknowledge that the techniques for creating a meaningful dialogue with the members of the public about nanotechnologies are still very much a work in progress (Pidgeon & Rogers-Hayden, 2007). Related approaches to upstream engagement include real-time technology assessment (Guston & Sarewitz, 2002) and upstream oversight (Kuzma & Tanji, 2010). These approaches typically have broader goals than developing participatory methods of public engagement. They are frameworks for guiding and evaluating the transition of an emerging technology into society, with upstream public engagement as one component. Yet the question remains: How should the task of engaging the public with nanotechnologies be approached?

Surveys that aim to elicit public attitudes toward nanotechnologies have started to proliferate (for an overview, see Satterfield, Kandlikar, Beaudrie, Conti, & Harthorn, 2009). But although these provide a useful snapshot of public opinion, they have only a limited ability to identify the social and ethical issues that novel technologies may raise (Pidgeon et al., 2005). As Fischhoff and Fischhoff (2002) have observed about biotechnology surveys, highly structured questions inevitably leave respondents guessing about the meaning of the questions and investigators guessing about the meanings of the answers. It is desirable to elicit *informed* rather than uninformed preferences—but providing information through survey materials is difficult. Surveys are often conducted over the telephone, or via the Internet, which makes the provision of more than a minimal amount of information problematic. But even when surveys are conducted face to face, researchers must be careful not to unintentionally frame the information they provide about a topic, and so information provision must either be kept to a minimum or carefully balanced. In order to provide a forum for nonexperts to think through the pros and cons of nanotechnologies (and their social and ethical implications), additional methods and approaches are required. Typically this means work that goes beyond asking survey questions and attempts to develop a model of engagement that is more participative.

In the first part of this chapter we will introduce the idea of upstream engagement, and present a short overview of pre-nanotechnology engagement work with a particular emphasis on agricultural biotechnology (i.e.,

GM crops). If, as it is often claimed, "lessons have been learned" about public engagement on emerging technologies from the GM experience, what are those lessons?

In the second part we introduce a framework for evaluating the efficacy of upstream engagement processes, posited by Kurath and Gisler (2009), and use this to examine existing upstream engagement work on nanotechnologies. Have the lessons learned from previous experience been applied—are there examples of nanotechnology public engagement that "work"?

In the third part of the chapter, we will foreshadow some of the issues that yet another upstream set of technologies raise—proposed attempts to geoengineer the earth's climate. It seems important that the lessons learned in engaging the public on nanotechnology are applied to upstream public engagement in other fields—and that lessons from other fields are utilized in nanotech initiatives. We conclude by presenting an argument for the continuing importance of upstream engagement, and the challenges for upstream engagement on nanotechnologies in the future. Despite the theoretical and methodological challenges associated with upstream engagement processes, promoting legitimate and participative public engagement with emerging technologies is an important goal—albeit one that researchers and policymakers must make more concerted and strenuous efforts to successfully achieve.

UPSTREAM ENGAGEMENT: LESSONS FOR NANOTECHNOLOGY FROM AGRICULTURAL BIOTECHNOLOGY?

The past decade has seen a concerted shift away from the "deficit model" of public engagement, which assumed that public opposition to science and technology was linked to a deficit of knowledge that could be addressed by public engagement (Hagendijk & Irwin, 2006; Irwin & Wynne, 1996; Renn, Webler, & Wiedemann, 1995). The deficit hypothesis has been discredited by empirical evidence—multiple studies have failed to find a straightforward link between a lack of knowledge about science and opposition to it (Sturgis & Allum, 2004). But the deficit approach has also fallen out of favor for another reason—it embodies the old-fashioned idea that public engagement is a one-way process, rather than a dialogue between scientists and the public. Although there is no universally agreed definition of public engagement, the Research Councils UK define public engagement as "any activity that engages the public with research, from science communication in science centres or festivals, to consultation, to public dialogue. Any good engagement activity should involve two-way aspects of listening and interaction."[2]

Public engagement can of course be pursued around any science or technology topic—but emerging areas of science such as nanotechnologies hold particular interest. This is because these technologies are "upstream," and as such are not yet subject to entrenched attitudes or social representations (Pidgeon & Rogers-Hayden, 2007). The motivations for upstream engagement are varied (see Fiorino, 1990; Renn et al., 1995; or Pidgeon, 1998, for discussion of some of the reasons for citizen participation in the development of a new technology; see also Pidgeon & Rogers-Hayden, 2007, for discussion in relation to nanotechnology), but a high-profile report from the British Royal Commission for Environmental Pollution (2008) emphasized that the full value of engagement and deliberation cannot be realized if engagement activities are (or are perceived as being) exercises in securing acquiescence to new technologies before they emerge (see also Dietz & Stern, 2008). Upstream engagement should not be interpreted as an opportunity for "resolving" concerns about a new technology at an earlier stage (Macnaghten, Kearnes, & Wynne, 2005). Instead, Stilgoe, Irwin, and Jones (2006) suggested that successful upstream engagement requires finding new ways of listening to and valuing diverse forms of public knowledge and social intelligence, and involving the public in more fundamental questions about the pace and direction of science and technology (see also Corner & Pidgeon, 2010; Wilsdon, Wynne, & Stilgoe, 2005).

However, whereas major dialogue and engagement processes have been used extensively in some other European nations such as Denmark, the Netherlands, or Switzerland (e.g., see Joss, 1998), until recently they have been less common in English-speaking nations—which is why engagement on nanotechnologies is seen as offering a potentially unique opportunity to practice or test upstream public engagement (Pidgeon & Rogers-Hayden, 2007). Public engagement activities occur all the time (and, for the most part, go undocumented), but interest in upstream engagement in Europe can be partly attributed to the widely held perception that public engagement over agricultural biotechnology (i.e., GM) resulted in something of a "backlash." Wilsdon and Willis (2004) argued that public engagement on GM began too late for public input to impact on research and development. Following this line of reasoning, in order to be considered legitimate upstream engagement must be conducted *before* major investment decisions have been made (Rogers-Hayden & Pidgeon, 2007).

Walls, Rogers-Hayden, Mohr, and O'Riordan (2005; see also Pidgeon et al., 2005) published a detailed review of three citizen-based deliberative procedures conducted in the UK, Australia, and New Zealand from the late 1990s to the early 2000s on GM. In the UK, the engagement program took the form of a series of public debates and focus groups known as "GM Nation." In Australia and New Zealand, consensus conferences were undertaken. According to Walls et al. (2005), in each of these three cases the stated aim was to broaden the public approval basis for a final political decision.

The British GM Nation events were organized by an independent steering group as a result of a recommendation to the UK government by the UK Agriculture and Environmental Biotechnology Commission (AEBC, 2001). The steering group of individuals held a range of views on the subject, and included industry and environmental NGO representatives. Due in part to substantial problems in reaching agreement over the stimulus materials to be presented to participants during the focus groups, many of the materials were prepared far too late to be sent to participants prior to their participation. They therefore had a very limited opportunity to engage with the evidence base, and experienced understandable difficulties in engaging in deliberative processes. Walls et al. (2005) described the public meeting as unstructured and poorly managed—producing significant barriers to participative debate (particularly for those participants who had not brought strong prior opinions to the table). Questions were also raised as to the representativeness of the focus group participants (Pidgeon et al., 2005).

Despite being organized successfully and receiving a positive appraisal from lay participants, the Australian consensus conference resulted in little or no impact on policymakers or policy. Walls et al. (2005) suggested that this was primarily due to the lack of any explicit commitments (prior to the initiation of the engagement process) that the outcomes of the consensus conference would be incorporated into future decision making. Subsequent reports from the Australian government made no mention of pursuing consensus conferences or deliberative strategies further in this field. In New Zealand, the structure of the engagement process meant that the consensus conference had a heavy technical focus that privileged a narrow conception of scientific risk over broader social and ethical concerns about the use and *application* of GM technologies. Therefore, despite the attempt at including public participation in the GM debate, the traditional science-policy approach taken by the New Zealand government ensured that very little convergence between the "pro" and "anti" sides was observed. Although consensus should not necessarily be the goal of public engagement, it is typically hoped that enhanced and more co-operative understanding will result (Dietz & Stern, 2008; Renn et al., 1995). Unfortunately, the New Zealand consensus conference did very little to ameliorate the animosity between opposing sides of the GM debate. From an inability to break from the traditional science/expert framing of the New Zealand project, to the significant barriers to participative debate encountered in the British initiative, Walls et al.'s (2005) conclusion was that "[a]ll three showed an inability to weigh explicit social value judgments with the broad science consensus . . . none of the(se) experiments contributed to a serious process of mutual social learning . . . The link to political procedures and to a wider array of public opinion was weak in all three cases."

It is frequently stated that lessons have been learned from experiences such as these. But what lessons are they? Kearnes, Grove-White, Macnaghten, Wilson, and Wynne (2006) examined exactly this question, observing that nanotechnologies are currently at a similar stage of development as agricultural biotechnology in the 1990s, and with similar levels of utopian promise, expectation and dystopian fear. Drawing on interviews with key individuals active in the GM debate, Kearnes et al. proposed that the lessons for nanotechnologies from the GM controversy are twofold—lessons about competing understandings of "science," and lessons about competing understandings of "the public."

Kearnes et al. (2006) identified several ways in which the GM public engagement experience might inform debates on nanotechnologies. First, the incredible optimism—almost evangelical—about the promise of GM technologies positioned it precariously: Anything short of fully realized potential would be seen as a failure. The "hype" around nanotechnologies is easily as great (cf. Berube, 2005). Although the passion and imagination of scientists and entrepreneurs are clearly a driving force in the development and ultimate acceptance of novel technologies, too large a gap between the rhetoric and the reality of nanotechnologies is likely to initially trigger public suspicion. From the perspective of developing an open and pragmatically oriented dialogue on nanotechnologies, it would seem to make sense to keep the hyperbole to a minimum. Kearnes et al. also suggested that the overly narrow focus on technical risks observed throughout the GM debate led to the exclusion of broader societal concerns. Already, there is evidence that (given the choice) members of the public choose to deliberate primarily about cultural and social rather than technical issues associated with nanotechnologies (Pidgeon, Harthorn, Bryant, & Rogers-Hayden, 2009).

In terms of ways of understanding the public, there are clear parallels between the approach frequently taken in debates about GM and the way that public engagement on nanotechnologies is developing. Kearnes et al. (2006) suggested that despite the discrediting of the deficit model, it has been resurrected in new forms in the nanotechnology case. Throughout the GM debate, public mistrust and skepticism were attributed to "mischievous" NGOs, or a sensationalist media. Although the public was not explicitly framed as "ignorant," skeptical views were frequently portrayed as arising from external factors like media exaggeration—rather than accepted as potentially legitimate dissent. Although public opinion toward nanotechnologies has not yet become dominated by a particular NGO campaign or strategy, Kearnes et al. argued that an increasingly inclusive approach to NGOs by policy and technology stakeholders is required—not least because the concerns they articulate often represent the sorts of broad, societal themes that a narrow focus on technical risk assessments can overlook.

A further cautionary note was sounded in a report produced by the Responsible Nano Forum (2009), which compiled reflections on the 2004 Royal Society and Royal Academy of Engineering Report (RS/RAEng) from experts in the field of nanotechnology and public engagement. A key element of the RS/RAEng report was its significant focus on the need for public engagement—leading to the depiction of nanotechnology as the "test bed" for public engagement (Pidgeon & Rogers-Hayden, 2007). Andrew Maynard (Responsible Nano Forum, 2009) observed that despite endless discussions, workshops, reviews, and reports on the responsible development of nanotechnology, few had taken the time to read and absorb the proposals in the report for "shifting gear" on public engagement. Although it has been argued that the *impact* of public engagement is often far harder to evaluate than the processes themselves (Bickerstaff, Lorenzoni, Jones, & Pidgeon, 2010), Maynard concluded (based on an assessment of expert perspectives from around the world) that evidence of meaningful change in regulatory practices and social engagement was hard to identify (Responsible Nano Forum, 2009).

In the following section, we review existing upstream engagement work on nanotechnologies. Are the lessons from GM being applied? And are there examples of public engagement on nanotechnologies that "work"?

NANOTECHNOLOGIES AND UPSTREAM ENGAGEMENT: ARE THE LESSONS BEING APPLIED?

Upstream engagement and nanotechnologies are closely related, so much so that, to reiterate, nanotechnologies have been described as the "test bed" for this form of public-science interaction (Rogers-Hayden & Pidgeon, 2007). The fact that nanotechnologies have acted as a test bed for upstream public engagement is important—because a great deal of interest has been paid to developing effective methods of engaging the public on nanotechnologies.

A growing number of public opinion surveys have been conducted asking questions about nanotechnologies (see Satterfield et al., 2009, for a review). This type of research is able to collect data from large numbers of people, and ensure that the participants are representative of a wider community, or population. Survey work is an important element of gauging public opinion on nanotechnologies, and it can focus on specific questions such as trust in regulators, attitudes toward different applications (Siegrist, Cousin, Kastenholz, & Wiek, 2007), and belief change on the basis of new information (Kahan, Braman, Slovic, Gastil, & Cohen, 2009). However, public opinion surveys do not on their own constitute public engagement. Public engagement requires—at a minimum—an attempt to develop a dialogue between science and nonscientists.

For some, interest in the public understanding of nanotechnology is geared toward identifying concerns at the earliest possible stage so that these concerns can be addressed in a timely fashion—because the successful development of nanotechnologies is dependent on public perceptions of their acceptability (Anderson, Wilkinson, Peterson, & Allan, 2009). A solid, upstream understanding of the public perception of nanotechnology allows research councils and stakeholders to stay one step ahead of the game—a fact that has not escaped the attention of industry stakeholders. In a 2008 report, Hart Associates observed that "the public's low level of familiarity with nanotechnology presents an opportunity for government, industry and the scientific community to establish confidence in nanotechnology-enabled products."

However, simply predicting and anticipating public responses to future applications of nanotechnology does not constitute satisfactory public engagement. Table 9.1 is a summary table of nanotechnology public engagement projects to date, although it is not designed to be exhaustive. In museums and science education centers across the world, small-scale public

Table 9.1 A List of Nanotechnology Public Engagement Projects, Ordered by the Date the Initiative Was Completed

Name	*Date*	*Location*	*Authors/ Organizers*	*Methodology*	*Output (Reports/ Publications/ Recommendations)*
1. Melbourne Citizens' Panel	2004	Australia	Commonwealth Scientific And Industrial Research Organisation (CSIRO)	One day expert/ community workshop	CSIRO report (Katz et al., 2005) reflecting on the results and implications for Australian policy
2. Royal Society workshop	2004	UK	Royal Society/ BRMB International Limited	Two exploratory workshops with lay participants	Reflection on results and methodology, plus recommendation for more comprehensive upstream engagement in RS/ RAEng (2004)
3. Informed Public Perceptions of Nanotechnologies and Trust in Government	2005	U.S.	Woodrow Wilson International Centre for Scholars	Three lay focus groups	Full report (Macoubrie, 2005) reflecting on results available at www.wilsoncentre.org
4. Nanojury UK	2005	West Yorkshire, UK	University of Cambridge; Newcastle University; Greenpeace: Guardian newspaper	Citizens' jury	Peer-reviewed publication discussing results and reflecting on methodology: Pidgeon & Rogers-Hayden (2007); "Perspectives" from witnesses, jurors, funders and facilitator available at http://www.nanojury.org.uk/perspectives.html
5. New Zealand Focus Groups	2005	New Zealand	The Agribusiness and Economics Research Unit (AERU) at Lincoln University	Series of lay focus groups	Full report (Cook & Fairweather, 2005) reflecting on results, methodology and implications available at www.lincoln.ac.nz/section165.htm

(continued)

Table 9.1 (continued)

Name	*Date*	*Location*	*Authors/ Organizers*	*Methodology*	*Output (Reports/ Publications/ Recommendations)*
6. Nanologue	2005/ 2006	Germany; UK	Wuppertal Institute for Climate, Environment and Energy; Forum for the Future; Triple Innova; EMPA	Dialogue and interviews	Three scenarios of how nanotechnology will have developed by 2015 and the Nanometer, an Internet-based tool assessing societal implications of nanotechnology (available at www.nanologue.net)
7. Small Talk	2005/ 2006	UK	Think Lab; University of Liverpool; British Science Association	Practitioner-led dialogue (including large debates and one-to-one conversations	Small Talk report (2006) reflecting on results of the project and the series of events that constituted it and drawing lessons for science communication.
8. Nanodialogues	2006	UK	Demos; BBSRC; EPSRC; Practical Action; Unilever; Environment Agency	Small-scale experiments; dialogue; workshop; focus group	Demos report (Stilgoe, 2007) discussing results and reflecting critically on methodology.
9. Publifocus	2006	Switzerland	Swiss Centre for Technology Assessment	Deliberative workshops/ focus groups	Peer-reviewed report discussing results: Burri and Bellucci (2007).
10. Governing at the Nanoscale	2006	UK	Demos; Lancaster University; ESRC	Expert interviews; public focus groups	Demos report (Kearnes et al., 2006) discussing results and reflecting critically on methodology.
11. Which? Citizens Panel	2007	UK	Which? and Opinion Leader	Citizens' panel	Full report reflecting on methodology and results is available at http://nanobio-raise.org/Members/susanne/news_item.2008-03-03.0873688285

12. Deliberating Nanotechnology in the U.S. and the UK	2007	U.S. (California); UK (Cardiff)	University of California Santa Barbara; Cardiff University	Comparative deliberative workshops	Peer reviewed publication (Pidgeon et al., 2008) reporting results and reflecting on methodology
13. Nanotechnology Grand Challenges: Healthcare	2008	UK	EPSRC	Structured decision analysis	EPSRC report (2006) reflecting on results available at http://www.epsrc.ac.uk/ResearchFunding/Programmes /Nano/RC/ReportPublicDialogueNanotech-Healthcare.htm
14. National Citizens' Technology Forum	2008	U.S. (various locations)	Centre for Nanotechnology in Society, Arizona State University	Six citizens' panels examining nanotechnology, biotechnology, information technologies, and cognitive science (NBIC)	Final reports for all six locations available athttp://www4.ncsu.edu/~pwhmds/final_reports.html. Each report contains the findings of the citizens' panel, and its recommendations. See also Hamlett, Cobb, and Guston (2008).
15. DEEPEN	2006–2009	Europe (various countries)		Deliberative for experts stakeholders and members of the public	DEEPEN report (2009) reflecting on results, methodologies and lessons for policy makers and funders of public engagement
16. South Carolina Citizens' School of Nanotechnology	ongoing	U.S.	South Carolina Citizens' School of Nanotechnology	A series of discussion sessions	Toumey, Reynolds, & Aggelopolou (2006) have reflected on ongoing role of South Carolina School in facilitating public engagement on nanotechnology

engagement events based on nanotechnologies are frequent occurrences (for a summary of some of these projects, see Gavelin, Wilson, & Doubleday, 2007).[3] However, many of these smaller-scale projects are not fully documented, and it is difficult to obtain reliable information about their design, methodology, and results. In many instances, their purpose is to simply provide a forum for public debate, rather than document results or methodology. We have therefore restricted the summary table to nanotechnology public engagement projects that have either been published in peer-reviewed academic journals or that have been documented by a full report reflecting on results, methodology, or both between 2004 and 2010.

Across the 16 nanotechnology public engagement projects listed in Table 9.1, there is considerable variation in the methodologies and approaches employed. Some are large-scale exercises that took place over a long period of time and involved large numbers of people (e.g., the three-year Small Talk project that took place in the UK, reaching 1,200 participants—see Project 6 in Table 9.1, Smallman & Nieman, 2006). Others are much smaller scale, such as the crossnational U.S./UK deliberation conducted by the University of Santa Barbara (see Project 12 in Table 9.1, Pidgeon et al., 2009). Besley, Kramer, Yao, and Toumey (2008) have questioned whether small-scale events are worthwhile with so few participants, but the number of people involved in a public engagement project is only one way of assessing its worth; it does not in and of itself tell us anything about the suitability of the methodology, the spread of opinion represented, the level of engagement and depth of data analysis or the seriousness with which the results were treated by policymakers and other decision makers.

Kurath and Gisler (2009) have offered a method of assessing the effectiveness of upstream nanotechnology engagement projects based on six criteria: the method employed, the framing of the science, the framing of the public, the style of communication, the impact on policy development, and the degree of self-evaluation. Although there is a great deal of work discussing best practice in public engagement (e.g., see Dietz & Stern, 2008; Renn et al., 1995; Rowe, Horlick-Jones, Walls, & Pidgeon, 2005; Sclove, 2010), Kurath and Gisler's criteria are aimed specifically at *upstream* engagement. According to Kurath and Gisler, an effective upstream engagement project is one that uses a participative methodology, avoids the traditional framing of the scientists as "experts" and the public as "nonexperts," uses a two-way dialogue style rather than a one-way communication of information from scientists to the public, has a measureable impact on policy or decision making, and reflects critically on its methodologies and results.

We have applied Kurath and Gisler's (2009) criteria to the projects listed in Table 9.1 as a means of evaluating their effectiveness. Some projects meet only a few, or none of the criteria. The Swiss Publifocus (Project 9 in Table 9.1), for example, took a fairly traditional approach to framing the science and the public in a series of focus group meetings. Information was presented by experts (a toxicologist and an ethicist), and the topics of discussion were specified by the organizers. The report on public opinion that was published based on its findings (Burri & Bellucci, 2007) did not reflect on the aims or methodologies of the project. Burri and Bellucci described the PubliFocus as being intended to contribute to public *awareness* and to act as an aid to decision makers. However, the aim of the project was not to develop a participatory dialogue with the public. This does not mean that the project was not worthwhile—simply that it was not a good example of upstream engagement—at least, not when judged by Kurath and Gisler's criteria.

Other projects in Table 9.1 meet some but not all of Kurath and Gisler's (2009) criteria. For example, the Melbourne Citizens' Panel on Nanotechnology (Project 1 in Table 9.1) utilized a format that involved "expert" presentations followed by questions from a relatively passive public—suggesting a deficit model approach to engagement. However, participants were then encouraged to form groups, to deliberate within a broad remit, and to formulate a group "position" on the hypothetical question of Australia's statement to the United Nations on nanotechnology. In this way, a two-way process of participation was encouraged, with the roles of expert and public reversed. The accompanying report (Katz, Lovel, Mee, & Solomon, 2004) also touched on the importance of promoting public engagement despite the risk of attitude polarization—that is, the authors took the characteristically upstream position that engagement has value independently of whether the public become more positive toward the topic in question.

An example of a U.S.-based public engagement project that fared quite well against Kurath and Gisler's (2009) criteria was reported by Macoubrie (2005, see Project 3 in Table 9.1). Four "experimental" groups were convened in three regional locations the summer of 2005, creating a total pool of 12 groups. The groups were experimental because they received different combinations of materials. The materials were designed to promote discussion rather than secure agreement, and participants were encouraged to respond in private (individually) as well as during group deliberation. However, the highly structured nature of the materials meant that participants' responses were focused quite narrowly on specific questions about physical risks and benefits, and the absence of information about these issues. In pre- and post-test questionnaires, participants were questioned directly about their trust in specific regulatory organizations, and Macoubrie's report included recommendations for increasing public trust in nanotechnologies—a normative goal not necessarily congruent with the criteria for best-practice upstream public engagement.

A number of the projects listed in Table 9.1 stand out as meeting all or most of the criteria specified by Kurath and Gisler (2009). Pidgeon and Rogers-Hayden (2007; see also Rogers-Hayden & Pidgeon, 2007) reported the findings of the UK Nanojury (Project 4 in Table 9.1). The jury had the opportunity to question "witnesses" (experts in the field) and were asked to reach a "verdict"—placing the public in control of the engagement process.

Pidgeon and Rogers-Hayden (2007) reported considerable difficulties in providing participants with a meaningful and realistic appreciation of the issues surrounding nanotechnology to *enable* them to act as jurors—some form of information provision is inevitable in upstream engagement work (a concern echoed by Flynn et al., 2009). However, because the report also reflected critically on methodologies employed, it provided a learning experience for the public and public-engagers alike. This is an important point for upstream engagement—if the methodology is not entirely successful, critical reflection can ensure that lessons can be learned for subsequent projects (a theme we develop in more detail in the final part of this chapter).

Pidgeon et al. (2009) used a deliberative workshop format to engage two quasi-representative groups of the public in the UK (Cardiff) and the United States (Santa Barbara) debating the risks and benefits of specific nanotechnology applications in two different domains (human health/enhancement, and energy—see Project 12 in Table 9.1). Although the purpose of the project was research, rather than engagement, participants were encouraged to deliberate open-ended and exploratory questions about the issues surrounding nanotechnology with only minimal "expert" input. The public and the science were therefore framed in an upstream way—without an assumption that the participants were there to passively receive information.

An example of a well-designed and successfully executed U.S.-based project was reported by Hamlett, Cobb, and Guston (2008—and see Project 14 in Table 9.1). Focusing on nanotechnologies for human enhancement, six citizens' panels were convened around the country, and recommendations from each location summarized in a report for the National Science Foundation (NSF). In a well-structured and carefully planned weekend of face-to-face deliberation, as well as a series of shorter online sessions, participants were offered a rich and comprehensive program of engagement. Participants drafted reports that represented the consensus of their local groups, and these local reports fed into an overall summary written specifically for the NSF—the leading funder of scientific research in the United States. Importantly, there was also an opportunity for participants to reflect on their experience of the engagement initiative itself—and the final report concluded that with the "appropriate information and access to experts, citizens are capable of generating thoughtful, informed, and deliberative analyses that deserve the attention of decision makers" (Hamlett et al., 2008).

The ambitious DEEPEN project (Davies, Macnaghten, & Kearnes, 2009—see Project 15 in Table 9.1) convened groups in the UK and Portugal, in addition to stakeholder consultations. The discussion groups were deliberative, but utilized novel and experimental methodologies for public engagement, including opportunities for participants to express their hopes and fears through performance or presentation. The project sought to move beyond the representation of public attitudes toward nanotechnology as "pro" or "anti," in order to establish the cultural narratives that characterized participants' responses to nanotechnology (the five narratives identified were 1) "be careful what you wish for"; 2) "opening Pandora's box"; 3) "messing with nature"; 4) "kept in the dark"; and 5) "the rich get richer and the poor get poorer").

Interestingly, however, a key recommendation of the DEEPEN project was that open-ended conversations on what nanotechnology may provide for society had only limited worth, and that concrete deliberation on possible developments in nanotechnology would be more productive for participants and researchers. According to the DEEPEN researchers, this is because identifying concerns about speculative futures makes the output of public engagement projects too easy for policymakers to ignore. Deliberating current or emerging research directions and extrapolating broader ethical issues from this base provide a potentially more powerful platform for engagement. In addition, the utilization of nonstandard methodologies such as performance and storytelling was designed to overcome the traditional "rational argument" bias of deliberation—which inevitably favors those participants who have mastered the communication skills associated with this style of interaction. The final report of the DEEPEN project was split into two sections, one of which was aimed specifically at policymakers, enhancing the likelihood of policymakers paying attention to the findings of the project.

The Nanodialogues (Project 8 in Table 9.1) consisted of four experiments in engagement, aimed at members of the public but also at stakeholders and research council members (including a "people's inquiry," a Zimbabwean workshop on nanotechnologies in water provision and two stakeholder events). The three-year program sought to explore not only public engagement with nanotechnologies, but also public involvement in research-council decision making—an essential component of ensuring that public engagement exercises are taken seriously by policymakers. The four experiments were broadly successful—the organizers described them as "proof of concept" that public engagement can make a positive difference. In terms of Kurath and Gislers' (2009) criteria, the experiments in the project consistently pursued a nontraditional framing of the science and the public, encouraged critical reflection from participants and evaluators, and made explicit attempts to create links between policymakers and public participants. From all sides there was enthusiasm for more (and better) public engagement.

However, the Nanodialogues team also reported that those in government who were willing to pay attention to public engagement exercises still tended to treat it as a way of building legitimacy—that is, they prejudged the outcome of public engagement by evaluating it on its capacity to increase support for or trust in technical decision making. The results of the first experiment were described as being an ineffective way for policymakers to gather evidence or to assess impact, but an effective lens through which policymakers could view the issues around nanotechnologies differently. The Nanodialogues report also made the important point that if upstream engagement is to be taken seriously, we cannot expect to observe the difference made by deliberative exercises immediately.

On the basis of our analysis, it would seem that there are several experiments in upstream nanotechnology engagement that have "worked"—that is, they have met most or all of the criteria developed by Kurath and Gisler (2009). Although identifying direct policy-level impact from public engagement exercises is notoriously difficult (Bickerstaff et al., 2010), a body of knowledge and experience now exists that can be meaningfully said to demonstrate effective and participative upstream public engagement. This is encouraging for the future of upstream public engagement—there are now precedents for how to successfully engage with the public, as well as documented pitfalls to avoid. The conclusion of the Responsible Nano Forum Report (2009) concurs with the analysis presented in the current chapter: It is possible to identify some good examples of upstream public engagement around nanotechnologies, but it would be inaccurate to claim that public engagement practices around nanotechnologies have dramatically shifted since the publication of the 2004 RS/RAEng report that called for renewed attempts at participative upstream engagement.

Although a comprehensive review and analysis of the *findings* of nanotechnology upstream engagement projects is beyond the scope of this chapter, it is instructive to briefly consider what some of outcomes of the projects listed in Table 9.1 have been so far, and to consider them alongside the results obtained from survey research. In a meta-analysis of 17 quantitative surveys conducted between 2002 and 2009, Satterfield et al. (2009) found that public familiarity with nanotechnologies is low, that people tend to see the benefits of nanotechnology as outweighing the risks, but that there is a great deal of uncertainty among participants in survey research with 44 percent of respondents claiming to be "unsure" about nanotechnologies. It is precisely this (sizeable) segment of the population that more deliberative and open-ended methodologies can attempt to reach through upstream engagement strategies.

Unsurprisingly, qualitative projects that administered "pre-engagement" questionnaires also found a low degree of awareness about nanotechnologies. For example, Macoubrie (2005—Project 3 in Table 9.1) reported that 54 percent of participants knew nothing about nanotechnologies prior to participating in the project. This lack of awareness rapidly disappears once participants have spent hours (or sometimes days) talking about nanotechnologies. It is more difficult to draw an "overview" conclusion from a body of qualitative research than it is to make a quantitative assessment of a series of surveys, but several deliberative projects have found significant positivity toward nanotechnologies. In the UK, participants in the Small Talk project (Project 7 in Table 9.1) highlighted some specific concerns, but were generally positive toward nanotechnologies (Smallman & Nieman, 2006). In the United States, Macoubrie (2005) found that a majority of participants felt the benefits of nanotechnologies outweighed their risks after several hours of deliberation. However, the conclusions from the projects listed in Table 9.1 do not always support the optimism that is typically expressed in surveys on nanotechnologies.

For example, Kearnes et al. (2006) found a latent ambivalence toward nanotechnologies that did not appear to diminish with greater knowledge and awareness. Several participants expressed an increased skepticism toward government and industry, and their ability to represent the public's interests regarding the regulation of nanotechnologies. Similarly, in the Nanodialogues report (Stilgoe, 2007, Experiment 4—see Project 7 in Table 9.1), participants expressed concerns about nanotechnologies that are difficult to capture using quantitative indices of risk and benefit. Although acknowledging that the benefits of a well-designed and honestly marketed product would ostensibly outweigh the risks in manufacture or consumption, extended period of deliberation led some participants to question the *need* for the product in the first place. This type of informed judgment bears on the *social context* in which science is conducted, rather than the science itself. Reflecting this distinction, Pidgeon et al. (2009) found that when given a choice between different aspects of the nanotechnology debate, both UK and U.S. participants preferred the social over the technical (Project 12 in Table 9.1). These outcomes do not necessarily represent a contradiction of surveys that find optimism toward nanotechnologies—but they do underscore the importance of the type of questions that are asked in determining the answers provided.

In the next section, we look ahead to the currently emerging topic of geoengineering. Can the lessons that have been learned be applied to other topics in the future? And what are the prospects for upstream engagement beyond nanotechnologies?

UPSTREAM ENGAGEMENT BEYOND NANOTECHNOLOGIES: PROPOSALS TO "GEOENGINEER" THE PLANET

Undoubtedly, nanotechnology stakeholders have learned from the debates around agricultural biotechnology that ignoring reasonable fears and concerns will slow (or possibly even prevent) progress. But based on our review of existing upstream engagement work, some more substantive lessons have also been learned about the philosophy and design of good quality participative engagement. Although we have not perhaps witnessed the paradigm shift away from the deficit model that some commentators have argued for, there is a greater degree of understanding that if public engagement is treated as a merely an opportunity for securing early acquiescence to nanotechnologies, the danger of a backlash is increased. However, it is important that the lessons learned from the GM debates and the tools that have been developed for studying nanotechnology are usefully applied to assess people's perceptions of the social and ethical implications of other upstream topics. One particularly pertinent example is geoengineering—proposals to engineer the earth's climate to prevent dangerous anthropogenic climatic change.

Geoengineering refers to the intentional manipulation of the earth's climate to counteract the effects of anthropogenic climate change. Proposals range from imitating trees' sequestration of carbon dioxide from the atmosphere by using giant chemical vents to "scrub" the atmosphere (analogous to the carbon capture and storage currently being developed for use on coal-fired power stations) to suggestions for the placement of trillions of tiny "sunshades" in orbit around the earth to deflect a percentage of solar radiation. Most of the technology implicated in geoengineering is at a pre-research and development phase, with no major research initiatives yet undertaken. But because geoengineering is beginning to be taken seriously by scientists around the world (including the American Meteorological Association and the British Royal Society), the technical, social, and ethical concerns about geoengineering will not remain upstream for long.

Corner and Pidgeon (2010) outlined what some of the pertinent social and ethical questions associated with geoengineering might be. There are salient issues around consent and conflict—who will decide whether geoengineering is necessary, and what will happen if global consensus cannot be achieved? There are also interesting questions around the way that geoengineering is being framed—as a "necessary evil" if current efforts at mitigation prove unsuccessful. Geoengineering the climate is a truly global proposition, and if the social and ethical questions it raises are to be satisfactorily addressed, then upstream public engagement on a global scale will be required. However, experience with nanotechnologies—which do not have the framing of urgency that proposals to geoengineer the climate so often do—suggests it will not be an easy challenge to meet.

Corner and Pidgeon (2010) argued that public engagement cannot be initiated after funding for technical research into the feasibility of geoengineering has begun, because by this point, consequential decisions about research funding have already been made. Responsible innovation and public engagement are critical for the oversight of research into emerging geoengineering technologies—a position articulated clearly in the "Oxford Principles" for governance of geoengineering research (Rayner et al., 2010). But relocating public debate about geoengineering to a sufficiently early point in its development will not address the generic difficulties of public engagement—who should participate, the efficacy of different approaches, and the need to ensure the results of any engagement exercise are taken on board by decision makers, as well as the means for locating dialogue within existing modes of democratic and public representation.

Like nanotechnologies, methods of upstream engagement on geoengineering must permit meaningful dialogue under conditions of high uncertainty and low awareness (what Funtowicz & Ravetz, 1992, have termed "post-normal" science). Typically, the participants in upstream geoengineering engagement processes will know little about the topic under consideration—and so information must be provided by the researchers. However, at a sufficiently early phase in the development of a technology, the "facts" are still contested (Rogers-Hayden & Pidgeon, 2007). How can upstream engagement provide information to the public, without unintentionally framing the issue?

Concerns such as these echo strongly the difficulties encountered in engaging the public on nanotechnologies. Upstream engagement brings a range of issues for the design of participatory exercises no matter what the topic. In the case of nanotechnologies the absence of products and easy everyday analogies through which people can interpret the science means they often struggle (initially at least) to get to grips with the topic. In turn this means that engagement mechanisms require at least some level of information provision about the issue, raising the question of how this information is framed and presented. In the case of geoengineering, many of the technologies currently being proposed may yet transpire to be little more than imaginative science fiction—but the social and ethical questions they raise warrant treating even the more fantastical proposals seriously.

Tracing a trajectory of public engagement from GM, through nanotechnologies and to yet-more upstream topics such as geoengineering highlights both the progress that has been made and the significant barriers yet to overcome for upstream engagement. The majority of the engagement projects reviewed in Table 9.1 demonstrate a lack of substantive self-evaluation or reflection (which, as Rowe et al., 2005, point out, is also a major challenge for more downstream public engagement). This may be attributable to the ever-present tension between policymakers' need for evidence and academic researchers' desire for impartial inquiry. Many of the projects detailed in Table 9.1 are collaborations between publicly accountable decision making or regulatory bodies and university departments. From a policy perspective, concrete evidence of policy impact and attitudinal shifts is a legitimate goal of public engagement. But Macnaghten (in Burchell & Holden, 2009) has argued that the role of an academic should not be to produce research that is of instrumental value to corporations or governmental bodies, but to critically engage in topical debate. Similarly, in the United States, there is a long and ongoing debate about the relative merits of "basic" and "applied" research—including the social and behavioral sciences (e.g., see NSF, 2007).

Concerns such as these speak to wider issues around the funding of university research. In the UK, an increasingly heavy emphasis has been placed on demonstrating the "impact" of academic investigation by the Research Councils, although this has proved unpopular among many university staff. It is possible that the shift toward impact-oriented research will not be conducive to promoting the sort of open-ended and reflexive methodologies that upstream engagement seems to require. However, these criticisms are arguments for improving and safeguarding the efficacy and legitimacy of upstream engagement—and indicate that it is difficult to match the theory of upstream public engagement with applied methodology.

Many attempts at upstream public engagement are characterized by a profound ambivalence on behalf of the public, accompanied by a general cynicism about whether the results of engagement exercises are taken seriously by decision makers and stakeholders. Clearly, without implementation at the policymaking level, public engagement will only ever have a limited impact. As the experience of GM engagement programs and the projects reviewed in Table 9.1 demonstrate, achieving adequate policy-level take-up and representation is a challenge. Decision makers must be careful not to use upstream public engagement as an opportunity to "get in early" with pro-technology public relations campaigns—but even more importantly, every effort must be made to make public engagement opportunities available to as many people (and as many different types of people) as possible.

In a good example of an attempt to broaden out public participation, a series of deliberations about climate change took place in 38 different countries (Danish Board of Technology, 2009). This represented an attempt to pursue public engagement on a global scale—a global citizen consultation involving 4,400 citizens. Each deliberation included around 100 participants, selected to be demographically representative for the region. Participants discussed and debated views on the policy goals of the United Nations Climate Change Negotiations in Copenhagen, with the results summarized for decision makers at those negotiations. Although 4,400 people constitute a tiny minority of the worldwide population, attempting public engagement on this scale demonstrates that public participation projects need not (and should not) be restricted to the citizens of industrialized, Western nations alone—a lesson that public engagement initiatives on any important topic would do well to heed.

CONCLUSION: UPSTREAM ENGAGEMENT ON NANOTECHNOLOGIES: THE CHALLENGES AHEAD

As a "test bed" for upstream public engagement (Pidgeon & Rogers Hayden, 2007), nanotechnologies offer a critical opportunity to learn from the experiences of the past and provide important guidance for public engagement in other upstream domains. As the literature reviewed in this chapter reveals, there are several examples of "good" upstream engagement on nanotech, as well as initiatives that have been less successful in meeting Kurath and Gisler's (2009) criteria for effective upstream public engagement.

Upstream engagement is critical because the uncertainties around nanotech are so high. Levels of scientific expertise and technical capacity are growing rapidly—but so is interest in the social and ethical questions that nanotechnologies raise. As the technologies societies develop become evermore powerful—in the case of nanotechnology, touching on questions about the very building blocks of life, and humans' relationship with nature—so the methods required to effectively vet and evaluate these technologies must develop too.

Upstream engagement processes are being refined using nanotechnology as the test case. But something notably absent from the upstream engagement debate is a clear sense of what public engagement is *for*—a confusion that stems primarily from the multiple motivations and perspectives of the actors and organization involved in engagement projects:

> Just as scientists have gravitated to "nanotechnology" without a clear idea of what it is, so people involved with "public engagement" have a multitude of meanings and motivations for doing it. We need to ask: "Public engagement as opposed to what?" Stakeholder engagement? Technocracy? Opacity? Authoritarianism? Public relations? Doing public engagement is fascinating, but it is not an end in itself. (Stilgoe, 2007, p. 75)

The continuing confusion over what public engagement projects are for is illustrated well by one final example of a UK nanotechnology public engagement project that was broadly seen as successful—Small Talk (Project 7 in Table 9.1). Smallman and Nieman (2006) summarized the multiple events that together comprised Small Talk—a program of nanotechnology public engagement that aimed to be of value to the public, science communicators, social researchers and policymakers. This was an ambitious and admirable goal. However, whereas the feedback from members of the public who attended the events (including focus groups, lectures, and debates) was positive, arguably policymakers stood to gain the most from the initiative—in particular from the "advice" on how to make nanotechnologies acceptable to the public. Although the Small Talk evaluation (Smallman & Nieman, 2006) also included plenty of pragmatic lessons for science communicators seeking to organize successful public engagement events (e.g., avoiding undue assumptions about prior levels of knowledge, being aware of the difficulty of discussing a technology with few concrete applications), the implicit assumption contained in this type of exercise—that nanotechnologies are acceptable and the problem is in getting this message across successfully—reflects a traditional view of the purpose of public engagement (i.e., as a form of market research). It is no trivial matter to transcend the competing (and often contradictory) aims that different agents in public engagement exercises have—policymakers want outputs, science communicators want clear procedures, business wants to develop markets, and the public want their voices to be heard.

Webler (1999) has described public participation in science and technology as an iterative development of "craft and theory," whereby academics (concerned primarily with research) develop theoretical principles of public engagement while heterogeneous practical programs provide experiential knowledge. Managing and reconciling these different components of public engagement will remain a challenge for those involved in designing upstream engagement projects around nanotechnologies (Flynn et al., 2009; Leach, Scoones, & Wynne, 2005).

However, despite the practical difficulties in successfully implementing upstream engagement, there is widespread agreement that it is beneficial for experts and policymakers to involve citizens in discussion about emerging areas of science and technology at the earliest possible stage. Nanotechnologies provide the perfect opportunity to turn this ambition into reality. In reviewing existing work and identifying successful examples of upstream engagement on nanotechnologies, we hope that this chapter contributes toward the important goal of pursuing participative and well-designed public engagement around nanotechnologies and beyond.

ACKNOWLEDGMENTS

This material is based in part upon work supported by the National Science Foundation under Cooperative Agreements Nos. SES 0531184 and SES 0938099 to the Center for Nanotechnology in Society at UCSB. Any opinions, findings, and conclusions or recommendations expressed in this material are those of the authors and do not necessarily reflect the views of the National Science Foundation. Additional support was provided by an Integrated Assessment of Geoengineering Proposals (IAGP) grant (EP/I014721/1) from the Natural Environment Research Council (NERC) and the Engineering and Physical Sciences Research Council (EPSRC).

NOTES

1. This is because many nanoparticles used in medical applications are engineered specifically to cross biological membrane barriers to assist drug delivery. It is precisely this novel characteristic, however, that is a cause for concern if they should enter the natural environment (Royal Commission on Environmental Pollution, 2008).
2. Although see also Rowe and Frewer (2005), who offer a detailed typology of public engagement mechanisms.
3. In the United States, for example, the Center for Nanotechnology in Society (CNS) at the University of Santa Barbara and the California Nanosystems Institute hold quarterly public engagement events where the public can question experts about nanotechnologies in an informal setting (http://www.cns.ucsb.edu/events-public-engagement-5/). In the UK, the science centre at Bristol hosted a series of events for young people to discuss the issues around nanotechnology and citizenship (http://www.at-bristol.org.uk/cz/Events/Default.htm#nanotechnology). In Denmark, the Danish Board of Technology conducted citizen interviews about public attitudes to nanotechnologies (http://www.tekno.dk/subpage.php3?article=1093&language=uk&category=11&toppic=kategori11).

REFERENCES

Agricultural and Environmental Biotechnology Commission. (2001). *Crops on trial*. London: Agricultural and Environmental Biotechnology Commission.

Allianz & OECD. (2005). *Opportunities and risks of nanotechnology*. Munich: Allianz.

Anderson, A., Peterson, A., Wilkinson, C., & Allan, S. (2009). *Nanotechnology, risk and communication*. Hampshire, UK: Palgrave Macmillan.

Bal, R., & Cozzens, S. (2008, September 24–26). *Public perceptions of NBIC technologies*. Paper presented at Prime Latin American Conference at Mexico City.

Berube, D. (2005). *Nano hype: The truth behind the nanotechnology buzz*. New York: Prometheus.

Besley, J., Kramer V. L., Yao, Q., & Toumey, C. (2008). Interpersonal discussion following citizen engagement about nanotechnology: What, if anything, do they say? *Science Communication, 30*, 209–235.

Bickerstaff, K., Lorenzoni, I., Jones, M., & Pidgeon, N. (2010). Locating scientific citizenship: The institutional contexts and cultures of public engagement. *Science Technology and Human Values, 35*(4), 474–500.

Burchell, K., & Holden, K. (Eds.). (2009). *The roles of social science in public dialogue on science and technology: Report of a one-day stakeholder workshop*. London: London School of Economics and Political Science.

Burri, R., & Bellucci, S. (2008). Public perception of nanotechnology. *Journal of Nanoparticle Research, 10*, 387–391.

Corner, A., & Pidgeon, N. (2010). Geoengineering the climate: The social and ethical implications. *Environment, 52*(1), 24–37.

Danish Board of Technology. (2009). *World wide views on global warming*. Retrieved March 2012, from http://www.wwviews.org/files/images/WWViews_info_sheet-v80-27_September%2009.pdf

Davies, S., Macnaghten P., & Kearnes, M. (Eds.). (2009). *Reconfiguring responsibility: Lessons for public policy* (Part 1 of the report on Deepening Debate on Nanotechnology). Durham, UK: Durham University.

Dietz, T., & Stern, P. C. (Eds.). (2008). *Public participation in environmental assessment and decision making*. Washington, DC: National Research Council, National Academies Press.

Drexler, E. (1986). *Engines of creation: The coming era of nanotechnology*. New York: Anchor Books.

Engineering and Physical Sciences Research Council, UK. (2008). *Report on public dialogue on nanotechnology for healthcare*. Retrieved March 2012, from http://www.epsrc.ac.uk/pages/searchresults.aspx?query=nanotechnology%20healthcare%20public%20dialogue

Ferrari, A., & Nordmann, A. (Eds.). (2009). *Reconfiguring responsibility: Lessons for nanoethics* (Part 2 of the report on Deepening Debate on Nanotechnology). Durham, UK: Durham University.

Feynman, R. (1959). There's plenty of room at the bottom. Retrieved March 2012, from www.its.caltech.edu/~feynman/plenty.html

Fiorino, D. J. (1990). Citizen participation and environmental risk: A survey of institutional mechanisms. *Science, Technology & Human Values, 15*, 226–243.

Fischhoff, B., & Fischhoff, I. (2002). Publics' opinions of biotechnologies. *AgBiotech Forum*, 4(3–4), 155–162.

Flagg, B. (2005). *Nanotechnology and the public: Part I of front-end analysis in support of nanoscale informal science education network*. Retrieved March 2012, from http://www.nisenet.org/sites/default/files_static/evaluation/NISE-FrtEndPart1Text.pdf

Flynn, R., Bellaby, P., & Ricci, M. (2009). *The limits of upstream engagement: citizens' panels and deliberation over hydrogen energy technologies.* Paper presented at the Society for Risk Analysis Conference: Karlstad, Sweden.

Funtowicz, S., & Ravetz, J. R. (1992). Three types of risk assessment and the emergence of post-normal science. In S. Krimsky & D. Golding (Eds.), *Social theories of risk*, pp. 251–274. Westport, CT: Praeger.

Gavelin, K., Wilson, R., & Doubleday, R. (2007). *Democratic technologies? The final report of the nanotechnology engagement group.* London: Involve.

Guston, D., & Sarewitz, D. (2002). Real-time technology assessment. *Technology in Society, 24,* 93–109.

Hagendijk, R., & Irwin, A. (2006). Public deliberation and governance: Engaging with science and technology in contemporary Europe. *Minerva, 44,* 167–184.

Hamlett, P., Cobb, M. D., & Guston, D. H. (2008). *National citizens' technology forum: Nanotechnologies and human enhancement* (CNS-ASU Report No. R08–0003). Center for Nanotechnology in Society, Arizona State University, Tempe, AZ. Retrieved March 2012, from http://cns.asu.edu/cns-library/type/?action=getfile&file=88§ion=lib

Hanssen, L., Walhout, B., & van Est, R. (2008). *Ten lessons for a nanodialogue: The Dutch debate about nanotechnology thus far.* The Hague: Rathenau Institute.

Hart, P. (2008). *Awareness of and attitudes toward nanotechnology and synthetic biology.* Washington, DC: Peter D. Hart Research Associates.

Irwin, A., & Wynne, B. (Eds.). (1996). *Misunderstanding science? The public reconstruction of science and technology.* Cambridge: Cambridge University Press.

Joss, S. (1998). Danish consensus conferences as a model of participatory technology assessment: An impact study of consensus conferences on Danish Parliament and Danish public debate. *Science and Public Policy, 25*(1), 2–22.

Kahan, D., Braman, D., Slovic, P., Gastil, J., & Cohen, G. (2009). Cultural cognition of the risks and benefits of nanotechnology. *Nature Nanotechnology, 4*(2), 87–90.

Katz, E., Lovel, R., Mee, W., & Solomon, F. (2004). *Citizens' panel on nanotechnology report to participants.* Melbourne, Australia: CSIRO Minerals.

Kearnes, M., Grove-White, R., Macnaghten, P., Wilsdon, J., & Wynne, B. (2006). From bio to nano: Learning lessons from the UK agricultural biotechnology controversy. *Science as Culture, 15*(4), 291–307.

Kearnes, M., Macnaghten, P., & Wilsdon, J. (2006). *Governing at the nanoscale: People, policies and emerging technologies.* London: Demos.

Kurath, M., & Gisler, P. (2009). Informing, involving or engaging? Science communication, in the ages of atom-, bio-, and nanotechnology. *Public Understanding of Science, 18*(5), 559–573.

Kuzma, J., & Tanji, T. (2010). Unpacking synthetic biology: Identification of oversight policy problems and options. *Regulation & Governance, 4,* 92–112.

Leach, M., Scoones, I., & Wynne, B. (2005). *Science and citizens.* London: Zed Books.

Macnaghten, P., Kearnes, M. B., & Wynne, B. (2005). Nanotechnology, governance and public deliberation: What role for the social sciences? *Science Communication, 27,* 268–291.

Miller, G., & Senjen, R. (2008). *Out of the laboratory and onto our plates: Nanotechnology in food & agriculture.* Australia: Friends of the Earth Australia Report.

National Science Foundation. (2007). *Merit review broader impacts criterion: Representative activities.* Washington, DC: National Science Foundation.

Nightingale, P., Smith, A., van Zwanenberg, P., Rafols, I., & Morgan, M. (2008). *Nanomaterial innovations systems: Their structure, dynamics and regulation.*

A report for the Royal Commission on Environmental Pollution. Retrieved March 2012 from http://www.rcep.org.uk

Oakdene, H. (2007). *Environmentally beneficial nanotechnologies.* UK: Department for Environment, Food and Rural Affairs (DEFRA). Retrieved March 2012, from http://archive.defra.gov.uk/environment/quality/nanotech/documents/envbeneficial-append.pdf

Pidgeon, N. (1998). Risk assessment, risk values and the social science programme: Why we do need risk perception research. *Reliability Engineering and System Safety, 59,* 5–15.

Pidgeon, N., Harthorn, B. H., Bryant, K., & Rogers-Hayden, Tee. (2009). Deliberating the risks of nanotechnologies for energy and health applications in the United States and United Kingdom. *Nature Nanotechnology, 4*(2), 95–98.

Pidgeon, N., Poortinga, W., Row, G., Horlick-Jones, T., Walls, J., & O'Riordan, T. (2005). Using surveys in public participation processes for risk decision making: The case of the 2003 British GM Nation? Public debate. *Risk Analysis, 25*(2), 467–479.

Pidgeon, N., & Rogers-Hayden, T. (2006). Reflecting upon the UK's citizens' jury on nanotechnologies: NanoJury UK. *Nanotechnology Law and Business, 2*(3), 167–178.

Pidgeon, N., & Rogers-Hayden, T. (2007). Opening up nanotechnology dialogue with the publics: Risk communication or 'upstream engagement'? *Health, Risk & Society, 9*(2), 191–210.

Rayner, S., Redgwell, C., Sauvulescu, J., Pidgeon, N., & Kruger, T. (2010). *Draft principles for the conduct of geoengineering research (the "Oxford Principles").* Reproduced in House of Commons Science and Technology Committee, The Regulation of Geoengineering, Fifth Report of the session 2009–2010, HC221.

Renn, O., & Roco M. (2006). *White paper on nanotechnology risk governance.* Geneva, Switzerland: International Risk Governance Council (IRGC).

Renn, O., Webler, T., & Wiedemann, P. (Eds.). (1995). *Fairness and competence in citizen participation: Evaluating models for environmental discourse.* Dordrecht, The Netherlands: Kluwer Academic Publishers.

Responsible Nano Forum. (2009). *A beacon or just a landmark? Reflections on the 2004 Royal Society/Royal Academy of Engineering Report: Nanoscience and nanotechnologies: Opportunities and uncertainties.* Retrieved March 2012, from http://www.responsiblenanoforum.org/pdf/beacon_or_landmark_report_rnf.pdf

Roco, M. (2004). The US National Nanotechnology Initiative after 3 years (2001–2003). *Journal of Nanoparticle Research, 6,* 1–10.

Roco, M., Mirkin, C.A., & Hersam, M.C. (Eds.). (2010). *Nanotechnology research directions for societal needs in 2020: Retrospective and outlook.* Boston: Springer.

Rogers-Hayden, T. and Pidgeon, N.F. (2007). Moving engagement "upstream"? Nanotechnologies and the Royal Society and Royal Academy of Engineering inquiry. *Public Understanding of Science* 16, 346–364.

Rowe, G., Horlick-Jones, T., Walls, J., & Pidgeon, N. (2005). Difficulties in evaluating public engagement initiatives: Reflections on an evaluation of the UK Gm Nation? Public debate about transgenic crops. *Public Understanding of Science, 14,* 331–352.

Royal Commission on Environmental Pollution. (2008). *Novel materials in the environment: The case of nanotechnology.* London: Twenty-seventh report of the Royal Commission on Environmental Pollution.

The Royal Society & the Royal Academy of Engineering. (2004). *Nanoscience and nanotechnologies: Opportunities and uncertainties.* London: Royal Society.

Satterfield, T., Kandlikar, M., Beaudrie, C.E.H., Conti, J., & Harthorn, B. H. (2009). Anticipating the perceived risk of nanotechnologies. *Nature Nanotechnology, 4*, 752–758.

Sclove, R. (2010). *Reinventing technology assessment: A 21st century model.* Washington, DC: Science and Technology Innovation Program, Woodrow Wilson International Center for Scholars.

Siegrist, M., Cousin, M. E., Kastenholz, H., & Wiek, A. (2007). Public acceptance of nanotechnology foods and food packaging: The influence of affect and trust. *Appetite, 49*, 459–466.

Smallman, M., & Nieman, A. (2006). *Small talk: Discussing nanotechnologies.* London: Think-Lab.

Stilgoe, J., Irwin, A. & Jones, K. (2006). *The received wisdom: opening up expert advice.* London, UK: Demos.

Stilgoe, J. (2006). *The received wisdom: Opening up expert advice.* London: Demos.

Stilgoe, J. (2007). *Nanodialogues: Experiments in public engagement with science.* London: Demos.

Sturgis, P., & Allum, N. (2004). Science in society: Re-evaluating the deficit model of public attitudes. *Public Understanding of Science, 13*, 55–74.

Toumey, C., Reynolds, J. R., & Aggelopolou, A. (2006). Dialogue on nanotech: The South Carolina Citizens' School of Nanotechnology. *Journal of Business Chemistry, 3*(5), 3–8.

Walls, J., Rogers-Hayden, T., Mohr, A., & O'Riordan, T. (2005). Seeking citizens' views on GM Crops—Experiences from the United Kingdom, Australia, and New Zealand. *Environment, 47*(7), 22–36.

Webler, T. (1999). The craft and theory of public participation: A dialectical process. *Journal of Risk Research, 2*(1), 55–71.

Werlin, R., Priester, J. H., Mielke, R. E., Kramer, S., Jackson, S., Stoimenov, P. K., Stucky, G. D., Cherr, G. N., Orias, E., & Holden, P. A. (2011). Biomagnification of cadmium selenide quantum dots in a simple experimental microbial food chain. *Nature Nanotechnology, 6*, 65–71.

Wilsdon, J., & Willis, R. (2004). *See-through science: Why public engagement needs to move upstream.* London: Demos.

Wilsdon, J., Wynne, B., & Stilgoe, J. (2005). *The public value of science: Or how to ensure that science really matters.* London: Demos.

Wood, S., Jones, R., & Geldart, A. (2007). *Nanotechnology: From the science to the social—the social, ethical and economic aspects of the debate.* London: Economic and Social Research Council.

10 Different Uses, Different Responses

Exploring Emergent Cultural Values Through Public Deliberation

Jennifer Rogers-Brown, Christine Shearer, Barbara Herr Harthorn, and Tyronne Martin

Emerging technologies such as nanotechnologies provide a unique opportunity to examine perceptions or judgments (Slovic, 1987) of technological risk and benefit as they are being socially and culturally produced, rather than retroactively. This chapter explores such "upstream" formative perceptions through six U.S. deliberative workshops conducted in California in 2009, informed by a prior 2007 Center for Nanotechnology in Society U.S./UK deliberation study. In both studies, participants deliberated the potential risks and benefits of nanotechnologies for either health/human enhancement or energy/the environment, to allow for controlled comparison of response to these different "application domains." We argue that, as participants interpreted workshop discussion points and materials, they drew on their understanding and values regarding health or environment/energy domains more broadly, including cultural knowledge about environmental and energy concerns, technological innovations, and the health industry, which impacted their risk and benefit perceptions of nanotechnology. In other words, application context affects public risk/benefit perceptions not because of the use of the technology per se, but because of the ideas and values tied to that application domain more broadly. The social issues of 1) urgency and necessity; 2) novelty; 3) regulation; 4) equitable distribution of benefits and risks; 5) privacy; and 6) responsibility are identified in this paper as particularly salient in influencing differing responses to application domain. These issues are operationalized as axes to allow for conceptualization of a multi-scalar model of emergent public responses to nanotechnologies.

WHAT CAN SURVEYS TELL US ABOUT PUBLIC PERCEPTIONS OF NANOTECHNOLOGIES?

In 2000, the U.S. government began funding research on nanotechnologies through the National Nanotechnology Initiative (NNI), and in 2003, the U.S. Congress explicitly included research on ethical, legal, and social issues (ELSI) as part of the NNI's funding scope. Because of the importance of public acceptance of new technologies and investment of public dollars in their development, one stream of nanotech societal dimensions or ELSI research has sought to explore and measure evolving public attitudes and perceptions about nanotechnology in general, including surveys.

A meta-analysis by CNS-UCSB researchers of all 22 published quantitative public surveys with the necessary data points for comparison from 2002 to 2009 in the United States, Canada, Europe, and Japan found ongoing low levels of public familiarity with generic "nanotechnology," with benefits viewed as outweighing risks by 3 to 1, but also a large (44.1 percent) minority who had not yet made up their minds about benefits or risks (Satterfield, Kandlikar, Beaudrie, Conti, & Harthorn, 2009). In public surveys, attitudes toward nanotechnologies have been found to be influenced by factors such as individual risk-benefit calculations (low concern for risks when benefits are high, and high concern for risks when benefits are low) (Currall, King, Lane, Madera, & Turner, 2006), cognitive shortcuts provided by the media (Scheufele & Lewenstein, 2005), pro-tech cultural views (Gaskell, Eyck, Jackson, & Veltri, 2005), and pro-tech individual views (Priest, 2006). Other surveys have focused more specifically on the influence of peoples' emotions and values, with Lee, Scheufele, and Lewenstein (2005) finding that affective reactions can override knowledge in the formation of risk/benefit perceptions of nanotechnologies, and Kahan (2008) arguing perceptions are informed by underlying individual values, making individual cultural and political dispositions key factors in the formation of risk/benefit perceptions, particularly for people with uninformed views about nanotechnologies (see also Kahan, Braman, Gastil, Slovic, & Mertz, 2007). More recent surveys have examined the influence of religion on views of technologies like nanotechnology that may be perceived as interfering with or violating nature (Scheufele, Corley, Shih, Dalrymple, & Ho, 2008; Vandermoere, Blanchemanche, Bieberstein, Marette, & Roosen, 2010).

As in risk perception and public attitude surveys more broadly, surveys of public views on nanotech have demonstrated demographic effects. A 2006 national survey by Peter Hart and associates found people reporting more perceived benefits than risks, although support was greater among men, young adults, and those making over $75,000/year, highlighting the influence of social location on views (Peter D. Hart Research, 2007).

Peter D. Hart Research's (2007) U.S. survey also suggested that support may vary by application: Despite reporting greater perceived benefits, only 12 percent of those surveyed said they would buy food packaging enhanced with nanotechnology, with only 7 percent saying they would buy nanotech food. Another question in the survey may shed light on the low support for food applications: When asked about the safety of the food supply, 61 percent said it has become somewhat or much less safe, whereas only 29 percent said it has become somewhat or much more safe. Further survey research on nanotech and food in Switzerland did indeed find public hesitation to buying nanotech food or food packaging, although in this study, too, food packaging was assessed as less problematic than food itself (Siegrist, Cousin, Kastenholz, & Wiek, 2007; Siegrist, Stampfli, Kastenholz, & Keller, 2008). Trust and high perceived benefits, as well as naturalness and perceived control, were strong factors in the willingness of those surveyed to buy nanotech foods and food packaging. Similarly, a 2008 telephone survey of 1,100 people found that although participant support for nanopills and nanofuels varied by how information about the application was presented (in terms of risk, controllability, bodily invasion, and social distribution), nanofood was uniformly unsupported, regardless of how information about it was conveyed (Conti, Satterfield, & Harthorn, 2011).

Other surveys have focused specifically on having participants rank benefits on the basis of application. In a representative national phone survey, Cobb and Macoubrie (2004) found peoples' preferred benefit (57.2 percent) for nanotechnology was "new and better ways to detect and treat human diseases." Cobb and Macoubrie gave their survey respondents five choices amongst "the most important benefits nanotechnology could achieve" and five "important potential risks nanotechnology should avoid." Besides treating human disease, the other preferred benefits included, "new ways to clean the environment," (15.8 percent), "increased national security and defense," (11.7 percent), "physical and mental improvements for humans," (11.5 percent), and "cheaper, better consumer products" (3.8 percent). Although participants ranked health applications higher than environmental benefits, 18.6 percent were concerned about health risks, specifically "breathing nano-particles that accumulate in the body." The highest ranked risk nanotech should avoid was a loss of personal privacy (31.9 percent). A separate 2005 study of experimental groups at three different U.S. sites asked the participants to identify up to five areas of highest interest regarding the benefits of nanotechnology, with 31 percent saying medical technologies, 27 percent greater consumer products, 12 percent individual and society advancement, 8 percent environmental protection, and 6 percent food and nutrition (the other options—economy/jobs, energy, computer/electronics, military uses/national security, and international welfare each got less than 5 percent) (Macoubrie, 2006). Yet Macoubrie also found that although U.S. public trust in government to manage the potential risks of nanotechnology was low, it was even lower when discussed in relation to medical uses, despite its rating as the highest area of interest regarding benefits. A 2008 survey of 556 people in the United States on human enhancement technologies also found high public support for nanotechnology applications perceived as improving human health (e.g., 88 percent approved of research on artificial vision for the blind) and detecting disease (84 percent), although there was low support for human enhancements like bionic technologies to increase strength (30 percent) (Hays, Miller, & Cobb, 2011). The same survey also showed that when respondents were asked with which applications they associate nanotechnology, machines and computers ranked highest (87 percent), followed by brain research (60 percent) and biological engineering (57 percent), whereas consumer products (47 percent) ranked lowest, suggesting a significant mismatch between public associations of nanotech applications and the actual nanotech applications currently on the market.

BEYOND SURVEY RESEARCH: WHAT DELIBERATIVE STUDY OFFERS

Surveys are offering insights into the potential effects of different applications on public views, with some assessing relative preferences or associations amidst a range of applications, whereas others have explored attitudes about a specific application, such as human enhancement or food. Significantly, these studies are increasingly suggesting that application context matters, affecting and shaping individual perceptions of benefit, risk, and importance (Pidgeon, Harthorn, Bryant, & Rogers-Hayden, 2009; Cacciatore, Scheufele, & Corley, 2011). Yet the structured nature of survey instruments can make it difficult to understand the underlying story and cultural/individual logics behind seemingly contradictory responses, such as high hopes for nanotech in medical applications and low trust in the medical and pharmaceutical industries. Further, how might individual associations of nanotechnology change or evolve through social interaction and discussion? Public deliberations help provide increased insight into emergent public views and responses to nanotech, while also engaging citizens as participants in discussions on the development of new technologies. We see our qualitative work, which also relies on some quantitative pre-/post-test data, as an expansion of previous survey work on public perception because public engagement allows for analysis of deliberative as opposed to rapid-fire intuitive judgments, enabling the tracking of emerging public views and the development of meanings/understandings of nanotechnology.

Deliberation research on nanotechnology/ies has a fairly extensive history already, based primarily in Europe. Due in part to the public backlash against GM foods, the UK government directed the Royal Society and Royal Academy of Engineering to carry out research on nanotech developments, leading to two public UK public discussion workshops in 2003. In response to the workshops, the Royal Society wrote a report arguing that public dialogue should be part of early UK nanotech development (Royal Society, 2004), helping lead to a series of UK public deliberations on nanotechnologies, including the 2005 five-week stakeholder session and public recommendations called Nanojury UK (Pidgeon & Rogers-Hayden, 2007), and a series of participatory events known as Small Talk in 2005 and 2006 (Smallman & Nieman, 2006). Nanologue, consisting of two months of interviews and discussions that took place in both the UK and Germany, led to an internet tool for people to offer their assessment of nanotech applications and products currently in development (www.nanologue.net). Deliberations were also held in Switzerland known as Publifocus (2006), where the focus was on discussion but expressly not on recommendations (Burri & Bellucci, 2008) and comparatively in the UK and Portugal, known as DEEPEN (2006–2009), which focused on nanotechnology and nanotech futures in general, which recommended that deliberations focus on concrete, ongoing developments rather than speculative scenarios that are easy for policymakers to dismiss (Davies, Macnaghten, & Kearnes, 2009).

Other deliberations have focused more on either soliciting general concerns and hopes with different nanotech applications, or focusing on public attitudes toward a specific application (such as nanotechnology in food production). In 2005, the Agribusiness and Economics Research Unit (AERU) in New Zealand used single session focus groups to measure views on nanotechnologies across different domains, specifically medical applications, computing/electronics, and food, and found agreement with both positive and negative statements associated with each application domain (Cook & Fairweather, 2006). Lancaster University and Demos conducted a series of public deliberation experiments called Nanodialogues in Swindon, England, with two of them focusing on specific application domains: one on nanotech for cleaning up contaminated land, and one on nanotechnology for cleaning water. Deliberating the use of nanotechnology for the treatment of contaminated land were London residents and a group of various experts (NGO, government, and university researchers), with the residents determining that such uses should not be applied until the long-term effects are studied and known and the results made public (Stilgoe, 2007). Three themes emerged from the focus groups: a precautionary approach to regulation, the need for more honesty and clarity from government and industry, and the importance of context for discussions about science, technology, and risk. In 2006 Zimbabwean residents participated in focus groups held by Demos, Practical Action, and the University of Lancaster, to discuss nanotechnology for cleaning water with scientists from around the world. The residents raised distributive justice concerns around access and accessibility (Grimshaw, Stilgoe, & Gudza, 2006). The 2008 National Citizens Technology Forum, undertaken by CNS-ASU researchers, convened citizen panels in six different states to get citizen input on human enhancement applications within the "converging technologies" of nanotech, biotech, information tech, and cognitive science (NBIC). Each panel gave a report listing their concerns, with all six sites concerned about the effectiveness of regulations for NBIC technologies and the need for public education, and five sites prioritizing funding for disease over enhancement (Hamlett, Cobb, & Guston, 2008).

A 2007 U.S.–UK cross-national study in the CNS at UC Santa Barbara (CNS-UCSB) took a more comparative approach to public deliberation, looking at how application can affect both the nature of individual and group responses and the form the dialogue takes. The cross-national deliberation study conducted by Barbara Herr Harthorn, and Karl Bryant in the United States and Nick Pidgeon and Tee Rogers-Hayden in the UK (Pidgeon et al., 2009) allowed controlled comparison between groups' views on nanotechnologies for energy and those for health and human enhancement. They found that participants in health/human enhancement and energy deliberative groups focused in general on benefits rather than risks of nanotechnology, yet support for nanotech applications for energy was more uniformly positive than for health and human enhancement technologies, which was more mixed and complex. Application context was, in fact, a far more salient difference than U.S.–UK national differences in contrasting group views concerning nanotechnology. The different group responses to health and energy applications seemed to be shaped in part by differing social/cultural issues tied to the different applications, such as perceptions about equitable distribution of benefits, government and industry trustworthiness, and risks to realizing benefits (Pidgeon et al., 2008).

Building on the 2007 U.S.–UK study, this current U.S.-based CNS-UCSB study was designed to allow a between-groups controlled comparison of views on applications to more explicitly explore these differing social/cultural issues and the impact of application context. The 2009 deliberations also varied by gender composition, thus allowing for between-groups comparison of group dynamics and perceived risks and benefits. Additionally, environmental applications were added to the energy deliberations.

Analysis of the 2009 U.S.-based CNS-UCSB deliberative workshops suggests that, although most participants leaned toward a benefits rather than risk frame, this was complicated by application context and did not necessarily fit along a single continuum of risks versus benefits; indeed, research has shown that in certain conditions, people may concurrently hold both high perceptions of risk and of benefit—these are not mutually exclusive attitudes or beliefs (Poortinga & Pidgeon, 2003). We argue that perceived risks and benefits are manifest primarily within a cluster of cultural/social concerns and issues associated with different technological applications, rather than individual calculations of technological risk per se (cf. Pidgeon et al., 2009). Because most people enter into public engagement knowing little to nothing about nanotechnologies, they often draw upon their hopes for scientific and technological advancements at the same time as they reflect on concerns with the industries in which nanotechnologies would be produced, shaping and informing their perceived risks and benefits around nanotechnologies.

In this chapter, we examine more closely how the specific application context within which nanotechnologies are given definition and shape produces distinctive discussions of risk and benefit. Informed by the 2007 deliberations, we explore how risk and benefit perceptions in these 2009 workshops differed for nanotechnology applications along six key contested axes concerning 1) urgency and necessity, 2) novelty, 3) regulation, 4) equitability, 5) privacy, and 6) responsibility.

METHODOLOGY

This qualitative research was designed to explore and develop new understandings of public expectations, values, beliefs, and perceptions regarding nanotechnologies through focused dialogue in deliberative public settings. Deliberative judgment is theorized by psychometric risk-perception researchers to provide a deeper set of understandings that may integrate cognitive and affective responses and so goes beyond the intuitive rapid judgment elicited in most survey designs. In this novel upstream research context, deliberative settings offer more methodological space than surveys to gain new understanding of emergent beliefs.

In July through October 2009, our interdisciplinary research team of social and physical scientists ran one pilot and six public deliberation workshops in central coastal California. Employing a between-groups comparative design (sometimes called a between-subject design in experimental social psychology; Davis & Bremner, 2006) the roughly half-day deliberative workshops focused on applications of nanotechnology for either energy and environment or health and human enhancement, with gender composition also varied systematically among the groups to include women only, men only and mixed gender groups. Each workshop consisted of a large focus-group size meeting in which we sequenced initial open discussion of the overt and covert cultural knowledges that surround the domains of technology, energy, and environment, or technology, health, and human enhancement, followed by carefully framed informational presentations about nanotechnologies generally and then about specific technologies within the application domain (Hill, 2005). For example, for the energy and environment workshops, we presented information about nanoporous membranes for water purification and carbon nanotubes for solar energy, and for the health and enhancement workshops we presented information about lab-on-a-chip nano-enabled devices and nanomaterials-enhanced prosthetics. This was followed by an open reading time, in which participants chose articles from a large range of relatively short written materials at varying levels of sophistication and detail on nanotechnologies, their relevant application domains, and potential implications. Participants read at least two different articles on their own, and then gathered into three smaller groups of three to five people, modeled after European style World Cafés, for minimally facilitated in-depth discussions (World Café Europe, 2010). The workshops concluded with a final hour-long dialogue with the full group.

Table 10.1 General Format of Workshop

1. Initial discussion of health/human enhancement or energy/environment
2. PowerPoint on nanotech basics and workshop theme (either nanotech for energy and the environment or health and human enhancement)
3. Working lunch: Participants read articles of their choice related to workshop theme
4. World Cafés: Three small group discussions: one on "nano basics" and two related to the workshop theme
5. Main discussion

Pre- and post-measures were used to provide a limited set of scaled judgments about risk and benefit perception, technological pessimism/optimism, societal/governmental responsibilities, and more open-ended thoughts and comments on public deliberation and these sessions. The workshops were audio and videotaped and these tapes were used to prepare full transcriptions of all dialogues. Data analysis is primarily qualitative via formal content and narrative analysis using NVivo software and systematic interpretive analyses by the team to determine key narrative themes and group dynamics. We worked inductively through in-depth readings and coding of the transcripts to identify themes and patterns as they emerged from group discussion. We employed a feminist, ethnographic content analysis that relied on reflexive analysis, discovery of emergent themes (including the axes discussed subsequently), and a broad set of categories to begin our "grounded" approach.[1] We also draw on quantitative analysis of the pre- and post-tests administered to all attendees to consider effects of participation in the workshops on the views held by participants.

The scientific information presented in the workshops and examples of specific nanotechnology applications were vetted with Nanoscale Science and Engineering (NSE) collaborators and NSE graduate researchers in the CNS and on the research team for accuracy and validity. In the energy and environment application sessions, examples focused on technologies for energy conservation (e.g., use of quantum dots in energy efficient lighting, LEDs; carbon nanotubes and engineered viruses for extending battery life), renewable energy (e.g., quantum dots for solar cells), and environmental remediation (e.g., carbon nanotubes for desalinating water; nanorust for filtering arsenic out of water). In the health and human enhancement application sessions, examples focused on medical technologies (e.g., lab-on-a-pill and lab-on-a-chip), and development of biomimicry nanomaterials to replace tissue and organs, targeted molecular drug delivery, nano "bombs" to explode cancer cells, and technologies that enhance human ability (e.g., enhanced prosthetics and memory enhancement).

Out of 67 total participants in six workshops, 33 were men and 34 were women, and each workshop ranged from 9 to 13 participants.

Recruitment and screening were performed by a professional third party (social science survey center), without disclosing the nanotechnology topic for the sessions. With our aim of democratic inclusiveness, recruitment was designed to produce a diverse quasi-representative quota sample of participants designed to match as closely as possible the demographics of the central coast California area along race/ethnicity, occupation/education, age, and income criteria, drawn from as diverse a set of recruitment points as possible. Following other deliberative research (Wilsdon & Willis, 2004), we held minimal education level to completion of high school.[2] A slight majority of the participants was White, 37/67 (55 percent); 15 (22 percent) were Latino; 6 (9 percent) were African-American; 5 (7.5 percent) were Asian-American; and 4 (6 percent) classified themselves as "Other." All participants were age 18 or over. A slight majority (55 percent) held a bachelor's degree or higher, 34 percent completed some college or an associate's degree, and 10 percent had only a high school degree. Age was fairly evenly distributed: 16 percent were 18–25 years old, 13 percent were 26–35, 15 percent were 36–45, 12 percent were 46–55, 21 percent were 56–65, and about 22 percent were 66 or older. A majority (58 percent) of participants reported a family income under $50,000, which is a bit high for Santa Barbara, where 48 percent of the population has an income of $50,000 or below.[3]

PRE-/POST-TEST RESULTS

Overall, our test results indicate a marked change in views about risks and benefits by participants from pre- to post-test conditions. As in the 2007 U.S.–UK CNS study, respondents at pre-test leaned toward a benefits over risk framework, with more than three to one claiming benefits will outweigh the risks, a quarter claiming the risks will equal the benefits, and 20 percent saying they were unsure. Results by application indicate that, at pre-test, a little over one third of participants in the energy/environment workshops reported that the benefits of nanotechnology will outweigh the risks, moving to almost half of participants at post-test, whereas health/human enhancement participants went from about one sixth benefits outweighing risks at pre-test to less than one third at post-test. More striking was the participant response to the question, "In the next 20 years, nanotechnologies will: [Improve our lives, Make them worse, or Make no difference either way]." At pre-test, the number of participants who thought nanotechnologies would improve their lives was comparable: 23 in the energy/environment workshops versus 18 in the health/human enhancements workshops. As indicated in Figure 10.1, by post-test participants in the energy/environment workshops, in spite of extended discussion on potential social and physical risks, overwhelmingly marked "improve our lives" (29/35 or 83 percent), whereas in the health/human enhancement workshop barely a majority (16/29 or 55 percent) affirmed this. This difference is due in part to the highly significant discrepancies between groups in the two applications in their "Not sure" responses (38 percent in health and 14 percent in energy), suggesting uncertainty was significantly greater in participants' minds (overall) at the end of the health/human enhancement workshops than the energy/environment workshops. These results are consistent with findings from the 2007 workshops, which showed greater unmitigated participant support for nanotechnologies used in alternative energy and energy conservation than for those in health and enhancement applications (Pidgeon et al., 2009), where social and distributive justice concerns in particular were found to affiliate strongly, alongside positive views about the benefits.

Why the different responses around risks and benefits and the future of nanotechnology in the energy/environment versus health/human enhancement deliberations? Further exploration of the differences evident in the quantitative survey data is made possible through contextual analysis of the workshop transcripts, in which we find individual and group perceptions of nanotechnologies shaped and complicated by social issues such as urgency, equitable distribution, and privacy, which differ in subtle and not-so-subtle ways across application domain. A similar set of themes was identified by the CNS-UCSB researchers in the 2007 workshops, and the protocol for the

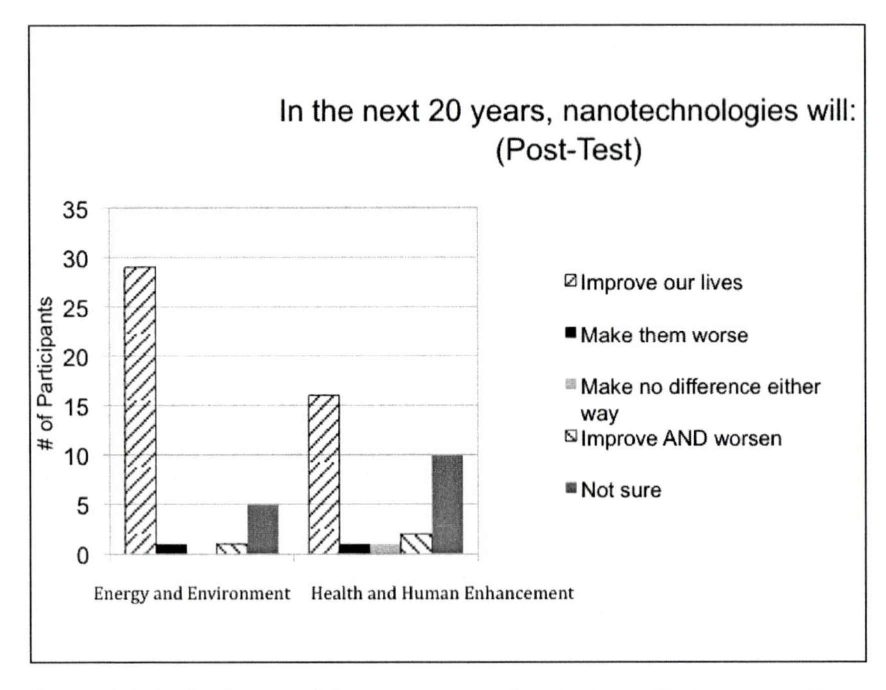

Figure 10.1 In the next 20 years, nanotechnologies will change our lives.

2009 deliberations was structured to more thoroughly explore these social issues and how they may differ by application context (technologies for health and human enhancement vs. energy and environment applications). These social issues, or themes, may be conceptualized as axes because the term signifies a continuum of possible responses and the ability for the axes to intersect and perhaps impact each other. The approach is similar to situational analysis (Clarke, 2005), which stresses the use of grounded theory to map out "major positions taken and not taken in the data vis-à-vis particular axes of difference, concern, and controversy around issues in the situation of inquiry" (p. xxii). The perceptions that lie along the axes are multi-scalar and therefore not dichotomous: Axes thus offer numerous possible responses or perceptions, not simply extremes at each end. The location of one response on an axis does not determine a response on another axis. There are, of course, an almost infinite number of potential axes. Our analysis focuses on those that recurred consistently across workshops and participants and captured key issues of central concern in the research.

These axes tell a story of emergent public opinions as they take shape through the social, linguistic, cultural, and cognitive processes involved in group deliberation about a set of issues. The next section explores participant discussion of each of these social issues in relation to the different application domains, and how responses to these issues co-vary with and impact risk and benefit perceptions.

KEY ISSUES FOR UNDERSTANDING RISK AND BENEFIT PERCEPTION: URGENCY AND NECESSITY

Although participants were excited about the potentials of nanotechnologies in both environment/energy and health/human enhancement domains, more of them conveyed a sense of urgency with regard to the energy/environment applications, and caution and hesitancy regarding health and especially human enhancement applications. Many expressed the idea that energy and environmental applications could help stave off global crises like climate change and dwindling water and oil supplies, making these applications seem both urgent and necessary. When the participants discussed an application as urgent, they saw the application as a necessary and important contribution to society and technology. In other words, the expressions of urgency were similar to those of necessity, except with additional affect, such as desire, need, and hope. A lack of necessity and urgency reflected a lack of interest in applications that seem wasteful of time and money or are potentially risky. We discuss subsequently examples of participant views of lack of necessity in relation to human enhancement applications.

Participants who expressed a sense of urgency for energy and environmental applications anticipated widespread personal and societal benefits, which overrode most concerns about potential risks. Surprisingly, even when participants expressed concern over risks and whether these applications would reach and benefit large numbers of people, they nevertheless still believed these technologies' full realization to be possible and highly important. The suppression (attenuation) of risk and concern has been noted to result from the perceived urgency for the development of a technology to solve a pressing societal problem (e.g., the oil or water crisis), making advocacy for more advanced technology (and trust in it) a form of risk attenuation (Kasperson & Kasperson, 1996).

Although participants were hopeful about and saw potential benefits in the use of nanotechnology for health and some human enhancement applications, these hopes often did not attenuate their sense of risk nor their moral concerns over issues such as cost, fairness, and distribution. Health applications like targeted cancer treatment and improved prosthetics were often discussed as highly desirable and necessary, but not necessarily urgent. This is interesting because few would doubt that, for example, cancer is an urgent medical problem. Participants' enthusiasm for nanotech health applications, however, was often immediately accompanied by moral concerns such as access, making many question whether such high tech treatment options would ever be available to everyone, even if immediately developed and marketed. Although nanotechnologies may offer participants new solutions for existing health problems, they would not necessarily offer new solutions for existing social problems like availability and access. Human enhancement applications, meanwhile, were often seen as desirable but largely unnecessary, although many participants conceded the lines between health and human enhancement applications are often blurry (e.g., improvements in hearing or vision) and therefore hard to differentiate in practice. Many also see human enhancements as already far advanced, and even technically radical potential nanotechnologies seem to appear as only incremental change in a direction already well established (if nonetheless a source of ambivalence).

The contrast between perceived urgency of nanotechnologies in the two main application domains is clear in the following quotes from workshop participants. In an energy/environment workshop, Lacy argues that concern over risk has never been a problem with technological innovation. In fact, she argues, risk management should not suppress nanotech research and development (R&D) that could improve the environment and solve resource depletion:

> 100–200 years ago, when they came up with new products, applications, whatever you want to call it, you know, we did not know any of the risks, but you can't stop improving your environment. You know, down the road I suppose it's hard to say, you know we don't know the effects that this is going to cause down the road, but it should not stop us from trying to do something. Right now we are running out of resources to continue the way we are going. I mean it is scary. (Lacy, Eng/Env F only, Woman, White, 46–55)[4]

For Lacy, it is much scarier to imagine running out of energy sources than to confront the hazards of potential future effects from nanotech development. The urgency of the moment, for Lacy, outweighs the unknown, potential future risks of nanotechnology. Further, Lacy's statement addresses a strong belief that technology can and should be used to address the problem of dwindling natural resources, and that it can do so adequately. Like many participants, Lacy expressed little doubt over the limits of technology to address energy and environmental problems, grounded within a robust faith in technological progress.

Lacy's comment highlights one of the key differences between the energy/environment and health/human enhancement workshops. Many of the participants in the energy workshops conveyed a sense of urgency surrounding the need for alternative energy and fuel to combat the growing reduction in global energy reserves. This sense of urgency about new technological development did not manifest to the same degree in the health workshops. The risk attenuation in Lacy's comment is related to the desire to dismiss risk because it cannot compare to the "scary" prospect of depleting global resources. What is notable about Lacy's comment (and those of many of her workshop partners) is the absence of any alternative solutions to the problem. She and many others in the groups have naturalized assumptions that people's (and societies') needs for energy are fixed, immutable, and growing, and she hopes that nanotechnology can solve all our energy problems, thus allowing us to continue our ever escalating rates of consumption. New energy technologies, however technically implausible (e.g., hydrogen fuel cells), are readily accepted by most of these U.S. participants as a panacea. In a similar vein, the implication of technology development (e.g., new IT) and the globalization of production in generating escalating demand for energy is little acknowledged or discussed.

Participants in the energy and environment applications workshops read and deliberated about water purification technologies. A majority of participants expressed excitement and a sense of urgency about these technologies. For example, Keith comments,

> I think it was that article that said only 1 percent approximately of the world's water supply is available at the right time at the right place, and so they said this can be the twenty-first century's petroleum, you know, could be the oil of the future. Since we're at war over oil right now *it's likely we could be at war over water in the future.* (Keith, Eng/Env M/F, Man, Asian, 56–65)

In this example, Keith expressed support for water purification technologies due to his concern that shortages in water pose an imminent threat and will lead to future global warfare, making the development and use of water applications urgent. Interestingly, the participants were generally pleased to learn about water purification programs in other countries, but often expressed dislike of programs in the United States (e.g., the possibility of water purification in the Hudson River by researchers who are testing the use of nano-zerovalent iron and nanopolladium in active carbon to aid PCB remediation), both because of uncertainty about risks to the broader ecosystem and because they tended to conceptualize clean drinking water as a more urgent and widespread problem in developing countries than in the United States (Lubick, 2009).

In contrast, we did not find health and human enhancement technologies discussed with the same degree of urgency, such as that evoked by impending environmental decline or war. Depending on the specific application (e.g., prosthetics, memory enhancement, or lab-on-a-pill) participants ranged in views from excitement to concern, but rarely spoke of urgency. Mary, for example, expressed her hopes for nanotechnology in health applications:

> I am the age my mom was [when she] came down with leukemia and she passed away a year later. And when I read these articles about . . . that they could go in and, and clear the white cells—act as a white cell seeking out the bad guys in the bloodstream, whether it is virus or bacteria or fungi or even cancer cells being able to identify them. I am just blown away with, you know, the possibilities. (Health/HE M/F, Woman, White, 66+)

Although hopeful and excited about the possibilities of nanotechnologies to fight disease, Mary does not express a sense of urgency about their development, despite her personal loss. This could be in part due to her social concerns around these technologies, which she expresses later in the workshop:

> I feared this at our other [discussion] table, but something that stuck out with me was that if we do not get this readily available to everyone, we are liable to create a society made up of two races—the wealthy nanoprotected or and then those that are not, are not able to access the possibilities.

Mary's statement reflects the belief that, even if nanotechnologies for health were developed, not everyone would get them, and these technologies may in fact increase existing social inequalities, arguably tempering her sense of urgency despite being "blown away by the possibilities."

Urgency was even less prevalent for applications viewed more as human enhancements. Gilbert, for example, expresses a common response to applications for human enhancement, especially applications that push the boundaries of human potential.

> I like the idea of small human functions, you know, getting improvement from your nanotechnology. Eyesight, hearing, even my complexion, that comes down to that, but the idea that you can turn me into a guy that can run fifty miles non-stop or swim to a battleship, holding my breath for half an hour, or make me live 120 years as a super being, [I] don't see the point in that. (Gilbert, Health/HE M only, Man, Latino, 56–65)

Although enthusiastic about improving certain human abilities, Gilbert expressed reservations about the need for its development, in contrast, for example, to the way people viewed nanotech applications for water purification. Further, his statement suggests there should be limits to what is developed in terms of health/human enhancements, with nanotechnologies that extend life or physical limits unnecessary, a line that was rarely articulated and drawn in relation to environment/energy applications.

Indeed, the difference between the health/human enhancement and energy/environment workshops is often further explained by analysis of the interplay between urgency and other concerns (axes) such as equitable distribution and trust (discussed subsequently). Mary's statement reveals one way the axes of urgency and equitable distribution may intersect. Although some excitement was expressed for the possibilities of nanotech R&D to lead to cures and treatment for disease, especially cancer, statements such as Mary's demonstrates how these hopeful comments were often mediated by a lack of trust in the equitable distribution of nano-enhanced health applications. As we explain in the "Equitable/Inequitable Distribution" section, participants in the health workshops believe that new health technologies will emerge just as other technologies are typically developed and distributed within the U.S. healthcare system, first to those with wealth and health insurance, and not necessarily to those who need them the most.

A common theme throughout all of the workshops, but particularly evident in the two all-women groups, was a prioritization of necessary and needed applications (products and technologies that may solve problems and benefit many) over "unnecessary" and redundant applications (technologies that already exist and do not justify additional unknown risks). As Gaskell et al. (2004) found with the acceptance of genetic modification, low perceived benefit of unnecessary nanotechnology applications results in less encouragement of the technology. Discussion of unnecessary (low benefit) technologies focused more on risk than those seen as necessary, urgent, and even "not urgent" but still somewhat beneficial applications. According to our participants, unnecessary nanotech applications could lead to more harm than benefit. Overall, alternative energy and water purification technologies were prioritized as the most beneficial and worthwhile, and thus necessary, reducing time spent on debate over risk. However, the health/human enhancement workshops had more discussion of risk and less sense of urgency, despite general benefit-oriented discussion and post-test survey responses.

EQUITABLE/INEQUITABLE DISTRIBUTION OF BENEFITS AND RISKS

Distributive justice includes concerns over equitable distribution of both the benefits and the potential risks of new technologies. Many participants questioned who would benefit from nanotech applications. Generally, people saw global benefit to be likely in the case of water applications, as they would relieve the lack of clean, accessible water in disadvantaged communities (which they tended to see as occurring external to their lives and the United States), but wondered about the potential of health and human enhancement applications to reach the broader public, including themselves.

Alberto weighed the potential risks and benefits of water purification and expressed the opinion that the benefits would be significant for the global world:

> I was pretty *positive* about it because I read in one of the reports that we just don't have a lot of freshwater available to us and it is going to become *more of a serious problem* depending on the other hand the toxicity levels everybody is talking about here today . . . but there is going to be *more people will benefit* and some people will probably suffer. (Alberto, Eng/Env M/F, Man, Latino, 26–35)

Participants in Alberto's workshop and the other energy/environment workshops generally believed that the potential benefits and perceived urgency of bringing clean water to those without access probably outweigh the uncertain risks. Additionally, we once again see an intersection between the axes of urgency and equitable distribution. In this example, Alberto sees more benefit than risk from water purification because of a sense of global urgency and need, despite awareness that this technology may not be accessible to everyone, and may cause toxic side effects.

Participants in the health/human enhancement workshops generally argued that nanotech health benefits would be distributed in the same manner healthcare is today within the United States—the wealthy and those with healthcare insurance are the first to benefit and gain access to new health technologies. The following exchange in an all-men health/human enhancement workshop illustrates this point:

> *Jordan (Latino, 18–25):* First off, the only people that are going to benefit are probably people with healthcare . . .
> *Gilbert (Latino, 56–65):* I think it'll be distributed, as you say, the same way it's distributed now. . .
> *Mitch (White, 36–45):* Yeah.
> *Gilbert:* . . . Benefits distributed now, the more money you have, the healthier you are. And that determines, you know, [the] access you have.
> *Mitch:* And the person who has a lot of money will be able to live a thousand years, and the people that don't, well . . .
> *Gilbert:* That one person who might be living a long time might be one of the richest.
> *Mitch:* They might want to keep that technology to themselves. (Health/HE M only)

These men clearly see a technocratic future where nanotechnology may solve problems, improve health, and perhaps help (some of) us live long beyond our current limitations. However, societal inequities seem likely to them to impede the benefits of this technology from being enjoyed by all such that access will probably be limited to the wealthy.

Jay, a participant in the human/health enhancement workshop, also expressed this concern:

> In the long term I find this [human enhancement technologies] tremendously exciting. I think that it would change our lives. So many of our dreams will come true. We might be able to—I mean just do things like flying and running really, really fast would just be amazing. I was discussing at the table that I think that one of the greatest wastes that there is, is that nature and time puts a great investment in people. So it is awful for people to die . . . and I think so long as this is available to everyone then that would remove one of the greatest tragedies of human existence. It is exciting so long as it is fair. (Jay, Health/HE M/F, Man, White, 36–45)

Jay is excited by the hope of extending life and was among a smaller subset of participants who saw death as a tragedy to be deferred as long as possible or avoided altogether, with nanotechnology as a potential route toward ending the tragedy. Jay, however, questions whether such novel developments will be "fair," and recognizes that unfair distribution would temper his enthusiasm over the potential benefits. Thus alternative energies were generally seen as accruing benefits at the societal level, with widespread recipients, yet health and human enhancement applications tended to be seen as more personal and individualistic, with many doubting that everyone will have access to such new health/human enhancement technologies.

NOVEL/NOT NEW

Nanotechnology is often held up as the "next big thing" and as a novel technology that will lead to revolutionary leaps in technological innovation. The typification of novel technologies as revolutionary, however, raises what Rayner (2004) terms "the novelty trap": debate over unknown consequences and the technocratic response that the new technology poses no new dangers—just as its predecessors. The novelty trap was evidenced in our deliberations, with participants generally expressing excitement about nanotechnology as a novel technology with endless possibilities, but simultaneously expressing concern over unknown (and potentially unknowable) risks. Consideration of these risks, however, generally came out more in the health/human enhancement deliberations than in those focused on energy/environment applications, clouding the enthusiasm over the ability of "new" technologies like nanotechnology to solve social problems.

The following quote from an all-women group shows a shared sense of excitement and hope for nanotechnology in new energy and energy conservation technologies, with the discussion evoking the novelty of nanotechnology:

> *Donna (White, 56–65 years):* I think it [nanotech for energy and conservation] is fabulous.
> *Facilitator:* Okay. Donna thinks they are fabulous.
> *Livia (Latino, 26–35):* Cutting edge.
> *Sara (White, 36–45):* Worth pursuing and continuing to research for sure.
> *Livia:* Endless possibilities!
> *Facilitator:* Endless possibilities.
> *Liberty (Other, 36–45):* Really cool!
> *Facilitator:* Really cool.
> *Liberty:* Buck Rogers. (Eng/Env F only)

This positive, technologically optimistic viewpoint was shared by many participants: "Technology historically has contributed enormously to improving standards of living and will continue it and in fact these technologies will even accelerate that trend" (Bert, Eng/Env M only, Man, White, 36–45). Bert's technocratic hope for the future of nanotechnology is based on a general belief in the ability of technology to improve society—it always has and always will.

Some participants, however, questioned the ability and limits of new technologies to solve old social ills, a feeling often expressed in the health/human enhancement workshops. Jordan, for example, believes health technologies will solve some problems, but that nothing can truly solve all problems:

> *Jordan (Latino, 18–25 years):* I think they're [health applications] exciting. I mean, but with all—but with the good there's going to be new bad, I guess.
> *Facilitator:* New bad?
> *Jordan:* Well yeah. I mean we're gonna have solutions for all these old problems but then new problems are gonna arise. You know it's like, it's like a never-ending cycle. We're never ever gonna get rid of famine and we're never gonna get rid of disease. It's something that's always going to pop up. (Health/HE M only)

For Jordan, nanotechnology may be the "new bad." Although nanotech could solve some problems, it could never solve the human fate of disease and death, and may even add new diseases or other new problems. This is in stark contrast to many energy/environment workshop participants who believed that new technologies can and should solve environmental and energy problems like dwindling resources.

Overall, the participants view nanotechnology as a novel approach to technological advancements. A few participants, however, questioned the newness of nanotechnology. For example, Jimmy, an Energy/Environment workshop participant said,

> I do not think I have seen anything where there was a totally new nanotechnology in its energy source, but from everything I saw it pretty much enhances what we are currently using now or makes it more efficient. (Eng/Env M only, Man, African-American, 56–65)

Jimmy's comment is similar to the views of science experts who also claim that nanotechnology simply offers incremental changes and improvements to existing technologies. However, our participants overwhelmingly argue the opposite, seeing nanotechnology as a novel technology with promising applications for both energy and the environment and health/human enhancement. This excitement for novel solutions and technologies, however, is tempered in discussions about health and human enhancement applications because of existing social concerns such as access, cost, safe regulation, and risk to human health.

NEED FOR REGULATION/NEED FOR FREEDOM FROM REGULATION

High perceived benefit (and low perceived risk) of emerging technologies is enhanced by trust in government, academia, and industry to advance technology safely, with shared benefits across publics. In contrast, expressions of high perceived risk (and low perceived benefit) often coincide with calls for greater regulation (Slovic, 1993). High perceived risk may emerge from technological pessimism and/or from lack of faith in corporations and government (recreancy) to manage the responsible development of technology (Freudenburg, 1993). Many participants expressed distrust in the willingness of corporations (and government) to prioritize public welfare over profits, raising concerns around exposure to potential risks. Further, with many participants seeing nanotechnology as a new technology, they wondered about potentially unknown risks: In the case of nanotechnologies, lack of research on toxicology combined with the novel characteristics of engineered nanoscale materials, which can, for example, pass through cell walls, was an often-cited point of concern in the workshops. Thus, with the exception of a very small group of participants who thought industry could regulate itself, government regulation and risk research were often seen as necessary to ensure public safety. This was true of both energy/environment and health/human enhancement applications, although health/human enhancement raised more concerns around direct harm to the human body and issues of privacy (discussed subsequently).

Many participants, for example, questioned the motivation behind corporations for the development of nanotechnology:

> I think that developing new technologies is great. I think one of the things that is missing the most is the motivation in terms of: What are you going to do with it? Are we going to exploit small countries? Is it for profit? Are we concerned about the waste matter later on? I think we are so want[ing] to develop that we don't think about what is this going to do. It is going to help us here? Is it going to hurt us here? (Oscar, Eng/Env M only, Man, African-American, 56–65)

Within the energy and environment workshops, the question of end-of-life waste issues was a consistent concern for many participants, like Oscar, who worries about corporations and government motivations as he wonders about the potential negative effects involved in the production of nanotech applications. He asks, "are *we* concerned" and later, "I think *we* are so want[ing] to develop that we don't think about what is this going to do." "*We*" could refer to the American public and/or corporations/industry/government. His alarm for what he feels is a general lack of concern about technological development and consequences highlights the problems associated with risk attenuation, apathy, and a drive for the faster, better, more profitable technologies: that "we" may get swept up in technological development and neglect consideration of possible risks.

For Tate, a participant in a health/human enhancement workshop, such drives for technological development means that regulation may come very late in the process:

> Well at this stage it's just an idea [regulation] because everything's so interesting and appears to be so beneficial. There's not gonna be much regulation because so many people are benefiting financially, but when the negative effects begin to, and enough people get hurt, and there's enough lawsuits, then the regulations will come. (Tate, Health/HE M only, Man, African-American, 46–55)

Similarly, other participants were concerned that nano-enhanced products were already on the market without their knowledge, raising issues around informed consent, labeling, education about nanotechnologies, and responsibility.

LOCUS OF RESPONSIBILITY

Issues concerning regulation also raise questions about responsible development of new technologies. We therefore asked the workshop participants about responsibility—who is/should be responsible for the environment, for individual health, and for safe and equitable distribution of nanotechnologies. Although the participants generally said that individuals and society are all involved in taking responsibility (for human health, energy conservation, and the environment), many believed that public laziness and consumerism are barriers to a shared sense of responsibility, especially regarding energy conservation. In all of the workshops, most participants claimed that industry, governments, and scientists are all responsible in varying degrees for safe development of nanotechnologies. The discussions of each application domain, however, also highlight different cultural values among health, energy, and the environment. Although the participants claim all individuals are and should be responsible for health and energy conservation, they produce judgments indicating they see more personal accountability for health maintenance (Callahan, 1994; Wikler, 2002).

With regard to informed consent and individual responsibility, many participants in the workshops generally claimed that it is the government's and industry's responsibility to educate people about nanotechnologies as a part of responsible development. Sarah comments,

> I think you need both [individuals and government]. I mean you have got the individual interplay, but you have got to have things like the EPA in place, in organizations, federally funded, to keep the checks and balances that listens to us, and that informs us, so we can do what we can do in our little lives, and they can do what they do on a larger scale. (Eng/Env F only, Woman, White, 36–45)

Sarah expresses some trust (or hope) in government to safely regulate and inform consumers about energy and environmental nanotechnologies. She sees the government playing a big role (at the "larger scale") and the public doing what they can with "our little lives," illustrating that many participants saw all actors as having responsibility over health, the environment, and safe regulation of nanotechnology. The question of personal responsibility versus community and governmental responsibility of human health is an ethical dilemma. Additionally, views on what is ethical (and risky) are impacted by attitudes about government/corporate/personal responsibility and trust in institutions, such as government, education, and industry (Siegrist & Cvetkovich, 2002; Slovic, 1999). The government has shifted positions over the education of health issues, taxation and insurance penalties for poor health choices, and the impact of socioeconomic status on longevity. Additionally, personal choice and actions, "rarely have all the attributes—informed, voluntary, uncoerced, spontaneous, deliberated, and so on—that in the ideal case, are preconditions for full personal responsibility" (Wikler, 2002).

Participants displayed the effects of the current devolution of societal responsibility onto individuals and expressed a particularly pronounced need for individuals to take on responsibility for human health because a lack of care is seen as detrimental not only to the individual, but also to families and communities through viruses and disease. If you are not responsible for your own health, then you hurt yourself *and* you may be hurting others by not treating a disease or illness:

> It is a societal thing. When you see back into the major scourges of the planet, yellow fever, malaria. Any kind of these things would, it's not about what you do for yourself but what your neighbor does and your surrounding society does in preventing that thing from spreading. So I think the best medical cure itself, healthcare, is really a societal concern because it affects all of us. If your neighbor becomes ill and can't afford to take care of himself, that stuff can impact you. (Gary, Health/HE M only, Man, White, 66+)

Additionally, participants placed responsibility for human health maintenance on charitable organizations and community groups in order to compensate for a lack of affordable healthcare:

> Well I think our foundations and charitable institutions have done a great deal. Some of them come up from behind the scenes that unfortunately in this economy are hampered, but I think traditionally in this country our tradition has been through charitable giving and all to support not only healthcare for those who can't afford it but also to assure that the availability of certain [care], even equipment is available to everybody that needs it and I think a combination of government and charitable giving is ideal. (Hope, Health/HE F only, Woman, White, 66+)

Regarding energy/environment issues, many participants expressed the view that if you do not take responsibility for saving resources and protecting the environment, you may be hurting your community, but not necessarily yourself (other than losing some money on your electric bill). This allows people to dismiss some responsibility with regard to energy/environment, particularly when they do not see others as taking on that responsibility. In the following exchange, participants in the men only energy/environment workshop discussed personal responsibility for energy conservation.

> *Oscar (African-American, 56–65):* Everybody has to do their part for the group.
>
> *Facilitator:* And then you have to restrain yourself? So you have to restrain the individual for the greater good. So, Alex?
>
> *Alex (Latino, 18–25):* I feel like, you know, like we can just conserving energy and stuff . . . will help us all, you know, feel better than like leaving running water and stuff. They just put it off.
>
> *Facilitator:* So do you think we have consensus in our society about that? As a sort of moral good? Bert?
>
> *Bert (White, 36–45):* You know a big chunk of our society is not responsible for anything.

Like most participants in the energy/environment workshops, these men expressed the belief that every individual *should* be responsible for issues such as energy conservation, but that most members of the public neglect that responsibility. Energy conservation helps the broader community, but does not provide a lot of personal gain, thus provoking calls in the group for monetary incentives for good behavior or technological fixes.

RISKY/NOT RISKY TO PRIVACY

Concerns over privacy were typically limited to health applications such as the lab-on-a-pill that would either pass through the human body or remain in a certain place within the body collecting information. Participants related their concerns to the current digital age, where private information is collected (and sometimes stolen) online. Gary's concern about privacy, for example, overlaps with his distrust of government:

> This always leads to other questions about what else does it do, reporting on what actions we're doing? Is it telling the police department or the FBI where I am at this time, I'm going over the Canadian border or what not. No, I'm not just making this up. (Gary, Health/HE M only, Man, White, 66+)

Many participants were concerned that employers and companies may intercept the information stored in the lab-on-a-pill in order to discriminate against them. Patty captures this concern well:

> What if you are applying for an insurance policy or something and they realize you have this preexisting condition or say you don't get a job because you have X thing that they don't want to hire you anymore because they don't think you are going to live more than 5 or 10 years. I mean the information can be used against you. (Patty, Health/HE F only, Woman, White, 46–55)

Patty's concern over intercepted data shows both belief in and acceptance of the broad possibilities of nanotechnology, and simultaneous distrust in the ability of industries to self-regulate and protect customers.

CONCLUSION

This study reaffirms the importance of considering application when assessing technology risk/benefit perceptions, and further explores *why* and *how* application matters by looking at six key issues to explore the cultural narratives produced around different application contexts. Many participants displayed starkly different reactions, for example, to nanotechnologies for solar energy and water purification than to nanotechnologies in medical and human enhancement applications. We explore how concerns about the risks and benefits of different nanotech applications are produced in dynamic interaction with social issues, such as urgency, novelty, regulation, equitable distribution, responsibility, and privacy. By focusing on these six key axes, we have shown how deliberative study reveals the complex and concurrent production of risk and benefit perceptions as they emerge in discussion of different nanotechnology applications. The intersecting axes reveal the dynamics of sense making, with effects on either risk amplification or increased benefit perception, depending on the application domain and convergent understanding across axes.

Our deliberations took place in the upstream, when public participants had little knowledge or familiarity with the physical, toxicological, or environmental risks of nanotechnologies, and therefore the participants articulate their concerns through their culturally constructed understandings of our current systems of development and distribution. Although their belief in the benefits of nanotechnologies is generally positive for health and even human enhancement nanotechnologies, their distrust in the fair, safe, and informed distribution of nanotechnologies by government and especially by industry and concern for unknown health risks mediate their excitement and hope for medical cures, treatments, and enhancements. Participants in the health and human enhancement deliberations expressed concern about access, informed consent, embodiment of nanoscale devices, and the fairness of enhancement technologies that may advantage some groups over others, tempering their hopes for health applications. In comparison to health/human enhancement, participants in the energy/environment workshops focused significantly more on the potential benefits of nanotechnology, such as the ability to solve urgent energy and water problems, and less often on questions about distribution, privacy, and health risks. The existence of both hopes and concerns suggests a level of upstream techno ambivalence (Harthorn, Shearer, & Rogers, 2011), but was more pronounced in the health/human enhancement workshops than those for energy/environment.

Early involvement with the public in dialogues about nanotechnologies offers the scientists, governments, and industries the opportunity to plan development with informed understanding of these differing public responses. The deliberations show something that our quantitative data cannot—the logics and associations through which participants digest the information and weigh the risks and benefits of each application that is presented to them, along a number of axes. As they deliberate, they necessarily situate the nanotech application within the social world, introducing into the discussion the fears, hopes, and general expectations they hold for themselves and members of society, such as the general public, scientists, politicians, and corporations, around these different technological domains. It is clear that the participants are conflicted about the benefits and risks of nanotechnology, and that "risk" and "benefit" may connote many different things to different people. Yet there were notable trends in individual responses along application lines. Exploring the cultural nuances of these emergent meanings will, we believe, offer a more accurate picture of the way people make sense of technologies, including new technologies like those at the nanoscale level. This, in turn, can help inform risk communication and risk management strategies that better understand and address the varying hopes and concerns that people bring to the citizen project of deliberating about different nanotech applications.

ACKNOWLEDGMENTS

This material is based upon work supported by the National Science Foundation under Cooperative Agreements Nos. SES 0531184 and SES 0938099 to the Center for Nanotechnology in Society at the University of California at Santa Barbara and NSF Grant No. SES 0824042 to coauthor Harthorn. Any opinions, findings, and conclusions or recommendations expressed in this material are those of the authors and do not necessarily reflect the views of the National Science Foundation. The authors would like to thank John Mohr for helpful comments on an earlier draft, student researchers Julie Whirlow, Rachel Cranfill, and Amanda Denes for their help with the transcriptions and data coding for this study, graduate student Indy Hurt for help piloting the workshops, and colleagues Nick Pidgeon, Terre Satterfield, and Karl Bryant, and for their many contributions to the wider project of which this is a part.

NOTES

1. For a classic example of ethnographic content analysis, see Atheide (1987).
2. For purposes of ethical research conduct and effective group interaction, including persons with less than a high school degree in activities that require ability to read texts and verbally discuss and debate ideas in English may be uncomfortable for them and counterproductive for the research (Jowett & O'Toole, 2006; Rosenberg, 2007).
3. Compared with the demographics of Santa Barbara, we slightly overrepresented for participants of color and people aged 18–25, and underrepresented for people making above $100,000. Participants were provided $75 for participation.
4. Notes describing a participant's social location and workshop participation are provided after each quote in the following format: pseudonym, workshop topic, gender composition of group, gender of participant, race/ethnicity of participant, and age of participant.

REFERENCES

Atheide, D. L. (1987). Ethnographic content analysis. *Qualitative Sociology, 10* (1), 65–77.

Burri, R. V., & Bellucci, S. (2008). Public perception of nanotechnology. *Journal of Nanoparticle Research, 10,* 387–391.

Cacciatore, M. A., Scheufele, D. A., & Corley, E. A. (2011). From enabling technology to applications: The evolution of risk perceptions about nanotechnology. *Public Understanding of Science, 20*(3), 385–404.

Callahan, D. (1994). Bioethics: Private choice and common good. *The Hastings Center Report, 24.*

Clarke, A.E. (2005). *Situational Analysis: Grounded theory after the postmodern turn.* Thousand Oaks, CA: Sage Publications.

Cobb, M. D., & Macoubrie, J. (2004). Public perceptions about nanotechnology: Risks, benefits and trust. *Journal of Nanoparticle Research, 6*(4), 395–405.

Conti, J., Satterfield, T., & Harthorn, B. H. (2011). Vulnerability and social justice as factors in emergent US nanotechnology risk perceptions. *Risk Analysis, 31*(11), 1734–1748.

Cook, A. J., & Fairweather, J. R. (2006). *Nanotechnology—ethical and social issues: Results from New Zealand focus groups.* Caterbury, New Zealand, The Agribusiness and Economics Research Unit, Lincoln University.

Currall, S. C., King, E. B., Lane, N., Madera, J., & Turner, S. (2006). What drives public acceptance of nanotechnology? *Nature Nanotechnology, 1*(3), 153–155.

Davies, S., Macnaghten, P., & Kearnes, M. (2009). *Reconfiguring responsibility: Lessons for public policy* (Part 1 of the report on Deepening Debate on Nanotechnology). Durham, UK: Durham University.

Davis, A., & Bremner, G. (2006). Research methods in psychology. In G. Breakwell, S. Hammond, C. Fife-Schaw, & J.A. Smith (Eds.), *The experimental method in psychology*, pp. 64–87. London, Sage.

Freudenburg, W. R. (1993). Risk and recreancy: Weber, the division of labor, and the rationality of risk perceptions. *Social Forces, 71*(4), 909–932.

Gaskell, G., Allum, N., Wagner, W., Kronberger, N., Torgersen, H., Hampel, J., & Bardes, J. (2004). GM foods and the misperception of risk perception. *Risk Analysis, 24*(1), 185–194.

Gaskell, G., Eyck, T. T., Jackson, J., & Veltri, G. (2005). Imagining nanotechnology: Cultural support for technological innovation in Europe and the United States. *Public Understanding of Science, 14,* 81–90.

Grimshaw, D.J., Stilgoe, J., & Gudza, L. (2006). The role of new technologies in potable water provision: A stakeholder workshop approach. *Report on the Nano-Dialogues held in Harare, Zimbabwe.* Practical Action.

Hamlett, P., Cobb, M. D., & Guston, D. H. (2008). National citizens technology forum: Nanotechnologies and human enhancement. *The Center for Nanotechnology in Society Arizona State University Report.*

Harthorn, B. H., Shearer, C., & Rogers, J. (2011). Exploring ambivalence: Techno-enthusiasm and skepticism in US nanotech deliberations. In T. B. Zülsdorf, C. Coenen, A. Ferrari, U. Fiedeler, C. Milburn, & M. Wienroth (Eds.), *Quantum engagements: Social reflections of nanoscience and emerging technologies*, pp. 75–89. Amsterdam: IOS Press.

Hays, S., Miller, C., & Cobb, M., (2011). National nanotechnology survey on enhancement results overview. In S. Hays, J. Robert, I. Bennett, & C. A. Miller (Eds.), *Yearbook of nanotechnology in society: vol. 3, Nanotechnology, the brain, and the future.* New York: Springer.

Hill, J. H. (2005). Finding culture in narrative. In N. Quinn (Ed.), *Finding culture in talk: A collection of methods*, pp. 157–202. New York: Palgrave Macmillan.

Jowett, M., & O'Toole, G. (2006). Focusing researchers' minds: Contrasting experiences of using focus groups in feminist qualitative research. *Qualitative Research, 6*(4), 453–472.

Kahan, D. M. (2008). Two conceptions of emotion in risk regulation. *University of Pennsylvania Law Review, 156*(3), 741–766.

Kahan, D. M., Braman, D., Gastil, J., Slovic, P., & Mertz, C. K. (2007). Culture and identity-protective cognition: Explaining the white-male effect in risk perception. *Journal of Empirical Legal Studies, 4*(3), 41.
Kasperson, R. E., & Kasperson, J. X. (1996). The social amplification and attenuation of risk. *The Annals of the American Academy of Political and Social Science, 545*, 95–105.
Lee, C., , Scheufele, D. A., & Lewenstein, B. V. (2005). Public attitudes toward emerging technologies: Examining the interactive effects of cognitions and effect on public attitudes toward nanotechnology. *Science Communication, 27*(2), 240–267.
Lubick, N. (2009). Cap and degrade: A reactive nanomaterial barrier also serves as a cleanup tool. *Environmental News, 43*(2), 235–235.
Macoubrie, J. (2006). Nanotechnology: Public concerns, reasoning and trust in government. *Public Understanding of Science, 15*(2), 221–241.
Peter D. Hart Research, Inc. (2007). Awareness of and attitudes toward nanotechnology and federal regulatory agencies. Washington, DC: Pew Charitable Trusts. Retrieved October 2011, from http://www.pewtrusts.org/our_work_report_detail.aspx?id=30539
Pidgeon, N., Harthorn, B. H., Bryant, K., & Rogers-Hayden, Tee. (2008). Deliberating the risks of nanotechnologies for energy and health applications in the United States and United Kingdom. *Nature Nanotechnology, 4*(2), 95–98.
Pidgeon, N., & Rogers-Hayden, T. (2007). Opening up nanotechnology dialogue with the publics: Risk communication or "upstream management"? *Health, Risk & Society, 9*(2), 191–210.
Poortinga, W., & Pidgeon, N. (2003). Exploring the dimensionality of trust in risk regulation. *Risk Analysis, 23*(5), 961–972.
Priest, S. (2006). The North American opinion climate for nanotechnology and its products: Opportunities and challenges. *Journal of Nanoparticle Research, 8*(5), 563–568.
Rayner, S. (2004). The novelty trap. *Industry and Higher Education, 6*, 349–355.
Rosenberg, S. W. (Ed.). (2007). *Deliberation, participation, and democracy.* Hampshire, UK: Palgrave Macmillillan.
Satterfield, T., Kandlikar, M., Beaudrie, C., Conti, J., & Harthorn, B. H. (2009). Anticipating the perceived risk of nanotechnologies. *Nature Nanotechnology, 4*(11), 752–758.
Scheufele, D. A., Corley, E. A., Shih, T., Dalrymple, K. E., & Ho, S. S. (2008). Religious beliefs and public attitudes toward nanotechnology in Europe and the United States. *Nature Nanotechnology, 4*(2), 91–94.
Scheufele, D. A., & Lewenstein, B. V. (2005). The public and nanotechnology: How citizens make sense of emerging technologies. *Journal of Nanoparticle Research, 7*(6), 659–667.
Siegrist, M., Cousin, M. E., Kastenholz, H., & Wiek, A. (2007). Public acceptance of nanotechnology foods and food packaging: The influence of affect and trust. *Appetite, 49*(2), 459–466.
Siegrist, M. & Cvetkovich, G. (2002). Better negative than positive? Evidence of a bias for negative information about possible health dangers. *Risk Analysis, 21*(1), 199–206.
Siegrist, M., Stampfli, N., Kastenholz, H., & Keller, C. (2008). Perceived risks and perceived benefits of different nanotechnology foods and nanotechnology food packaging. *Appetite, 51*(2), 283–290.
Slovic, P. (1987). Perception of risk. *American Association of the Advancement of Science, 236*(4799), 280–285.
Slovic, P. (1993). Perceived risk, trust, and democracy. *Risk Analysis, 13*(6), 8.

Slovic, P. (1999). Trust, emotion, sex, politics, and science: Surveying the risk-assessment battlefield. *Risk Analysis, 19*(4), 689–701.

Smallman, M., & Nieman, A. (2006). *Discussing nanotechnologies: Small talk report*. Retrieved October 2011, from www.smalltalk.org.uk/downloads/small_talk_final_report.pdf

Stilgoe, J. (2007). *Nanodialogues: Experiments in public engagement with science*. London: Demos.

Vandermoere, F., Blanchemanche, S., Bieberstein, A., Marette, S., & Roosen, J. (2010). The morality of attitudes toward nanotechnology: About God, techno-scientific progress, and interfering with nature. *Journal of Nanoparticle Research, 12*(2), 373–381.

Wikler, D. (2002). Personal and social responsibility for health. *Ethics & International Affairs, 16*(2), 47–55.

Wilsdon, J., & Willis, R. (2004). *See-through science: Why public engagement needs to move upstream*. London: Demos.

11 News Media Frame Novel Technologies in a Familiar Way

Nanotechnology, Applications, and Progress

Erica Lively, Meredith Conroy, David A. Weaver, and Bruce Bimber

In an era of rapid technological and social innovation, those who are typically outside the development of these innovations—the public—confront a flood of new technologies with many societal consequences, few of which can be anticipated much in advance. Through individual choices and social processes conditioned by culture, through market forces, and through the actions of the state, the public absorbs or occasionally rejects new technologies. Over the last two decades, much of the attention to social effects of innovation has focused on technologies of information and communication, which clearly define many aspects of the contemporary period. However, other technologies, especially those broadly termed "nanotechnology," also present important possibilities for social change. Nanotechnology is less deeply diffused into daily life, and far less salient publicly, but it has the potential for lasting consequences. Nanotechnologies present the public with potentially novel issues of risk and uncertainty, combined with disparate benefits for many groups of citizens. They have already been the target of a considerable investment of public funds.

These facts raise interesting questions for those interested in a forward-looking view of technological innovation as a social force. At this early stage, it is a good time to ask what Americans think about nanotechnologies, and just as importantly, to ask *how* they think about innovation. People may worry that advances in nanotechnology could lead to new kinds of toxins or environmental hazards, which could shape the trajectory of regulation and consumer markets in one direction. Or people might think about nanotechnologies in terms of new and better ways to treat diseases. or they might come to covet a better set of golf clubs, fortified with nano-coating, that are more resilient and ostensibly offer more "power." These could contribute to a different trajectory for markets and the involvement of the state. It is also possible that people may simply come to view new nanotechnologies as an inevitable and desirable product of scientific advance, rather than a subject for politics.

The term "nanotechnology" covers a wide range of materials, devices, industrial sectors, and applications, although nanotechnologies are at this stage only now emerging into the marketplace. Nanotechnology is employed in biomedicine, health and human enhancement, information technologies, consumer goods, energy conversion and conservation, and food safety and storage, as well as other areas. Although most proponents of research and development have focused on nanotechnology's immense potential and considered it a key group of technologies for the early twenty-first century, scientists and developers of nanotechnologies are not in unanimous agreement regarding the environmental, and health and safety impacts of nanomaterials. Uncertainty about potential risks, combined with low levels of public knowledge, suggest the potential for a wide diversity of public opinion. Survey research shows that people perceive the benefits of nanotechnology to outweigh the risks (Pidgeon, Harthorn, Bryant, & Rogers-Hayden, 2009; Scheufele & Lewenstein, 2005). Yet the majority of respondents in surveys know little to nothing about nanotechnology (Cobb & Macoubrie, 2004; Currall, King, Lane, Madera, & Turner, 2006; Gaskell, Eyck, Jackson, & Veltri, 2005; Satterfield, Kandlikar, Beaudrie, Conti, & Harthorn, 2009). This suggests the potential for uncertain shifts in future opinion that may hinge on whether news and public discourse about nanotechnology emphasize risks, rewards or both (Zaller, 1992).

Unlike technologies of information and communication—which are often tangible and familiar, and are to a considerable degree under the control or partial control of people who use them—nanotechnology is different. It cannot be seen, and it is almost always embedded in other products or tools. Moreover, unlike some products made with nanotechnology—such as computer parts or an improved tennis racquet—engineered nanoparticles have the potential to be used directly in interaction with human biology in medicine and related disciplines. Perceptions of nanotechnology will likely matter more than reality in shaping people's behavior, and perceptions will in turn be shaped powerfully by news coverage—in particular by how media frame stories related to nanotechnology. A "frame" is the way that a news narrative emphasizes certain aspects of a story over another, in such a way as to potentially influence what values or judgments are evoked in the minds of the audience (Entman, 1993). Whether frames as vivid as "death tax," "tax relief," or "Frankenfood" will ever be associated with an application of nanotechnology remains to be seen. In the long run, the presence and salience of vivid, evocative nano-associated frames will likely be a function of the actions of advocacy groups, the advertising efforts of corporations, the shifting attention of the news system, position-taking by high profile public officials and, of course, developments in labs and high tech markets.

Although it is too soon to know what future approaches to news about nanotechnology may be, scholars do know that media coverage shapes public perceptions of difficult concepts with generalizations and often simplistic frames, and that nanotechnology is no exception (Brossard, Scheufele, Kim, & Lewenstein, 2008; Nisbet et al., 2002; Scheufele & Lewenstein, 2005). In this chapter, we explore news media coverage of nanotechnology, examining what it looks like across time, news outlets, and applications. One of our chief questions concerns the extent to which news media frame nanotechnology in terms of "progress." It is well known that news media tend to report on science and technology in terms that are positive, futuristic, and that invoke ideas of a natural order of "progress" (Nisbet et al., 2002). Viewing technology through a lens of progress can de-politicize it, directing attention away from the interests that drive innovation, its distributional consequences, and the socially constructed nature of technology (Bijker, Hughes, & Pinch, 1987; Ellul, 1967; Winner, 1977). We are interested in the extent to which the "progress" frame is used by news media. Nanotechnology provides a potentially illuminating case in which to see how frames become associated with a novel set of technologies for which journalists do not have storytelling experience.

In exploring coverage of nanotechnology, we build on previous work in which we analyzed the prevalence of media frames in newspaper coverage of nano (Weaver, Lively, & Bimber, 2009). Here we update that work with new data, and extend the analysis further. We address four main questions regarding the form and content of media stories.

1. How has the volume of print news stories focusing on nanotechnology changed over time?
2. What is the frequency of print media use of nanotechnology frames over time?
3. To what extent do newspapers frame unique nanotechnology applications differently?
4. Do different media outlets tell distinctly different stories about nanotechnology?

These questions are in service of a larger theme of interest to us, namely, how media coverage of technology can either clarify or obscure the extent to which such technological change involves the exercise of political power. Depoliticized discourse about technological innovation has long been recognized as a threat to the accountability of corporations as well as government for their powerful roles in social changes (Ellul, 1967). In what follows, we review some of the key principles of framing effects generally. Following that brief review, we describe recent findings from the framing literature specifically with regard to nanotechnology in media. Lastly, we address our research questions through a content analysis of print media coverage of nanotechnology from 1999 through 2009. We conclude with a discussion of our findings.

FRAMING EFFECTS

In general, media have the potential to influence and, under some circumstances, change individuals' expressed preferences and opinions. Media influence works through several mechanisms, from influencing which considerations or thoughts are most readily accessible "off the top of the head" (Zaller, 1992), to the priming of deeply held convictions or values that people draw upon to make decisions or judgments (Entman, 1993). Under the best circumstances, news presents filtered information that makes sense of unfamiliar or controversial topics, reducing the complexity of decision making. For example, stem cells were at the center of controversy in 2001 when President George W. Bush limited federal funding for research using embryonic stem cells. Prior to this decision, public debate centered on two arguments supported by news: first, that the government has an ethical imperative to restrict funding for researchers who pursue science at the expense of human embryos, which some believe are human life; and second, that the government has an ethical imperative to fund stem cell researchers who are pursuing cures for debilitating diseases and to save human lives (Nisbet, Brossard, & Kroepsch, 2003). These arguments simplified what was, in actuality, a series of interlocking scientific, ethical, legal, and moral considerations. By presenting two basic arguments, the media largely enabled members of the public to quickly choose a side based on minimal elaboration, as well as facilitating the application of their views on a related issue, abortion.

Framing is "the process by which people develop a particular conceptualization of an issue or reorient their thinking about an issue," and framing effects occur when changes in the presentation of an issue produce a change in opinion (Chong & Druckman, 2007, p. 105; see also Ajzen & Fishbein, 1980). Frames have a well-documented influence on individuals' attitudes and opinions about public policy issues (Chong & Druckman, 2007; Haider-Markel & Joslyn, 2001; Jacoby, 2000; Nelson, Clawson, & Oxley, 1997). "Emphasis frames," or "issue frames" (Druckman, 2001) are so called because they highlight certain aspects of an issue over others. A classic demonstration of the effect tested the influence of emphasis frames on attitudes toward the Ku Klux Klan in the laboratory. One frame was more sympathetic to a KKK rally, emphasizing members' right to assemble and freedom of speech. The other frame emphasized the possibility of violence and threats to social order. Subjects exposed to the frame emphasizing the First Amendment expressed greater tolerance for the Klan than did those exposed to the public order frame that highlighted the potential for a clash (Nelson et al., 1997).

As Nelson and Oxley (1999) observe, "[F]rames are more than simply positions or arguments about an issue. Frames are constructions of the issue: they spell out the essence of the problem, suggest how it should be thought about, and may go so far as to recommend what (if anything) should be done" (p. 1057). A frame affects the way individuals perceive a problem and its consequences, and thus possibly alters their evaluation of the issue (Slothuus, 2008). As Entman (1993) noted in his classic study, "[T]o frame is to select some aspects of a perceived reality and make them more salient" (p. 52). The underlying theory is that the temporary accessibility of ideas and concepts induced by media frames—although tempered by chronic differences among individuals' dispositions and tendencies in processing information differently—influence our ideas, attitudes, and opinions on important issues. The fact that framing effects extend beyond the question of whether people are "for or against" an issue such as stem cell research is crucial. Frames can affect how people think about the causes and remedies of problems and which of their own values are implicated.

Iyengar (1991) found that describing a social problem such as unemployment in broad, thematic terms led people to attribute blame to institutions or societies. However, exposure to episodic reports of unemployment in specific terms with concrete stories, or exemplars, led people to blame the unemployed themselves for their plight. In this way, how people might respond to policy proposals for social welfare programs, or perhaps regulation of new technologies, may be tied to framing approaches that may not themselves contain any explicit position about policy or responses.

Although frame analysis has most often been employed in analyzing explicitly political issues, it has also been useful in studying public perception of scientific topics. In the case of nanotechnology, political as well as technical developments in the 2000s presented journalists with an increasing set of options for framing news. Stories in the English-language press now numbers in the thousands (Friedman & Egolf, 2005; Weaver & Bimber, 2008). Scheufele and Lewenstein (2005) and others (e.g., Brossard et al., 2008) find support for the hypothesis that media frames play an important role in shaping individuals' attitudes in the United States toward nanotechnologies, specifically. These authors report that attention to mass media is one of the strongest predictors of attitudes toward nanotechnology. News media are then critical in shaping opinions.

FOUR FRAMES FOR TECHNOLOGY

In our previous work, we identified four prominent frames in nanotechnology news coverage: progress, generic risk, regulation, and conflict. We found these four frames to capture virtually all stories about nanotechnology (Weaver et al., 2009). We use these frames again for the present analysis. The *progress* frame links technological change to a "natural" process of scientific discovery and to improving social conditions. It minimizes issues of responsibility, choice, priorities, and regulation. Not unlike Iyengar's (1991) episodic frames, the progress frame in effect says that technology just happens. For example *The Houston Chronicle* wrote,

> Nanotechnology enables humans to build things from the ground up. It involves a world so small researchers believe that at some time in the future, nanotechnology can be used to fix human cells that are

> diseased, heal wounds in minutes or repair damaged hearts. They envision nanotechnology devices that could be built to kill only cancer cells without destroying healthy tissue. ("Nanotechnology," 2004, n.p.)

Notice that in this passage the *Chronicle* assigns a kind of agency to technology, which enables unnamed humans to accomplish heroic goals. Nano exists in a special world in which researchers discover things necessary for curing cancer. Notice also what is missing from this passage: Just a partial list would include the fact that taxes likely fund the research being described, rather than going toward other goals; that corporations stand to profit from this new world; that agencies exist with responsibility for regulating new products for the safety of the public; that researchers themselves are not in agreement about the safety of all nanotechnologies or whether any particular benefits will come to pass.

The *generic* risk frame focuses on possible harms or potential problems associated with nanotechnologies. It calls attention to the fact that there may be a price to pay for nanotechologies' benefits, but in this frame, the author stops short of calling for regulation or oversight as a remedy. For example, in 2001, the *Houston Chronicle* ran a story with the headline, "Nanotech Encounters New Barriers: Environmental Risks Rise as Costs Decline," with a focus on the possible negative environmental effects of nanotech. This article quotes the late Rice University professor Richard Smalley as saying, "This is the Darth Vader side of nanotechnology … it is conceivable that some variance of the nanotechnology we will make will have a very nasty future. There will be all sorts of applications in nature" (Berger, 2001, p. 31). Both the headline and the quote highlight the negative impact of nanotechnologies and go so far as to link nanotech with a readily accessible, through fictive fear: an image of Darth Vader, the once valiant Jedi who inevitably falls to the dark side in the film *Star Wars*. The generic risk frame is one step more complex than the progress frame, opening the door to political questions, but it stops well short of identifying people, interests, corporations, or government agencies who are players in the development of technologies. The risks or "dark side" are intrinsic to technology, just as are the promised benefits, rather than arising from the actions of researchers or corporations. At its worst, the generic risk frame could imply that one can not have progress without paying some costs.

The *regulation* frame departs from the progress and generic risk frame by introducing the political process explicitly though abstractly. It emphasizes the necessity of government action to protect the public, especially when nanotechnology is used in production of consumer products. For example, on January 11, 2006, *The San Jose Mercury News* ran a report by the Associated Press (AP) titled "Safety Laws Trail Behind Nanotech Development, Report Says," which opens by stating that, "laws protecting the public's health and safety aren't developing nearly as quickly [as the technology]" (Bridges, 2006). News articles revolving around the regulation frame largely highlight the risks of nanotech as just cause for regulations. Another example comes from *The Washington Post*:

> Lab animal studies have already shown that some carbon nanospheres and nanotubes behave differently than conventional ultrafine particles, causing fatal inflammation in the lungs of rodents, organ damage in fish and death in ecologically important aquatic organisms and soil-dwelling bacteria. An estimated 700 types of nanomaterials are being manufactured at about 800 facilities in this country alone, prompting several federal agencies to focus seriously on nano safety. Yet no agency has developed safety rules specific to nanomaterials. (Weiss, 2005, p. A08).

The *Post* article stresses the rationale and need for regulation, without necessarily suggesting ways to overcome the regulatory lag. The psychometric risk literature has shown that statements of the need for regulation produce amplified risk perceptions (Slovic, 2000), so stories with a regulatory frame are likely to affect readers' perceptions both of the political process and the technology itself. The AP article aptly concludes, "if there are not adequate safeguards, the public is going to resist the technology and it won't meet its potential" (Bridges, 2006).

Last, the *conflict* frame in stories of technology highlights disputes between specific societal interests, actors, and competing claims. The conflict frame presents nanotechnology issues in concretely political terms. The *New York Times*' article titled "Nanotechnology Has Arrived: A Serious Opposition Is Forming" is an example. In the article, the author writes that the early days of nanotech's discovery—in which conflict was limited to a relatively small community of nano researchers and experts largely fighting about novel risks—are over. Today, more mainstream organizations and players now take up the fight: As consumer products such as sunscreens and popular clothing incorporate nanotechnology, consumer safety advocacy organizations have a more obvious role. For example, the ETC Group has called for a moratorium on self-assembly and self-replication, while other groups similarly call for slower incorporation of nanotech into our everyday lives, at least until the risks are more certain (Feder, 2002). Groups such as the Science and Environmental Health Network usually find themselves up against nanotech scientists, though news articles adopting the conflict frame are unlikely to personify the conflict as a fight between scientist and consumer. Instead, it is usually portrayed as a fight between consumer and environmental organizations.

We previously found that these four frames captured essentially all stories about nanotechnology in English-language print news media around the globe from 1999 to 2008 (Weaver et al., 2009). Yet we found that no single frame or pair of frames emerged to dominate the coverage, suggesting journalists' conception of "nanotechnology" was either ambivalent or multivalent.

Given the near certainty of broad application of nanotechnologies in the future, our present analysis aims to explore whether different domains of nanotechnology application receive more coverage and are subject to specific frames. For example, nanotechnology as applied to health might receive more coverage in terms of risk (or benefit), whereas nanotechnology as applied to information technology might receive more coverage in terms of progress (for an analogical example of this in the area of biotechnology, see Marks et al., 2007). This rationale is grounded in findings from survey research showing differential responses across application: this entails more positive attitudes toward energy applications than health applications and generally negative reactions to food applications (Pidgeon et al., 2009; Rogers, Shearer, & Harthorn, 2011; Siegrist, Cousin, Kastenholz, & Wiek, 2007). We suspect that media coverage may also exhibit varying approaches and frames across application areas. It is not clear theoretically whether these should align with public reactions, which are likely shaped both by media coverage and by qualities of the technological applications themselves, such as their invasiveness with respect to the body.

To examine these issues and answer our four main questions, we look at print coverage of nanotechnology from 1999 to 2009 in 12 major publications. We divided applications of nanotechnology into nine major domains: human health, medicine, consumer goods, food, energy, computers and information technology, environment, economy, and a miscellaneous category. We arrived at this set of categories by adapting the major application areas advertised by the National Nanotechnology Initiative itself.[1] We also look at the volume of print coverage of nanotech over time more generally.

NEWS STORIES

The news outlets of interest for this study are the top U.S. newspapers by circulation. We included the nine largest-circulation papers that are available on LexisNexis: *USA Today*, *Wall Street Journal*, *New York Times*, *Los Angeles Times*, *Denver Post*, *Washington Post*, *New York Daily News*, *New York Post*, and *Houston Chronicle*. These papers are important for several reasons. They include the major newspapers of record in the United States, and are the primary newspapers in several major cities. They also supply stories to affiliated news businesses or are part of common distribution networks of stories through parent organizations, especially Gannett, McClatchy, and the Tribune Company, which together operate more than a hundred newspapers. We conducted full text searches on articles in these papers, with the exception of *Wall Street Journal*, for which LexisNexis only provides abstracts. Additionally, we included the *San Francisco Chronicle* and the *San Jose Mercury News* because they represent a region that produces large amounts of technology-centered news. We also included the Associated Press, because it is an important supplier of "wire-service" stories to a great many news businesses both large and small (Weaver & Bimber, 2008).

The search criterion we used to populate the sample of stories about nanotechnology was any permutation of the word *nano* ("nano!"). We chose this simple search criterion in order to obtain as complete and thorough a list of news stories as possible. Our goal of capturing every story relating to nanotechnology did not call for sophisticated techniques for automatically discriminating among kinds of stories or categorizing results. We manually screened out irrelevant stories, such as those about the Apple iPod Nano and those containing the term "nanosecond." The resulting stories contain traditional terms such as nano, nanoscience, or nanotechnology as well less frequently used words such as nanoparticle, nanomedicine, or nanotube. Once the search results were returned, we then read every title and lead paragraph in order to identify those with a lead frame predominantly about nanotechnology.

This search criterion differs from that in Weaver et al. (2009). The present sample includes all news, including stories with no particular attention to societal issues; in the earlier work, we filtered out news with no coverage of any societal issues, in order to reduce the bias toward progress frames. The current sample is therefore larger and more inclusive in that it captures all news about nanotechnology and not just specific news about the societal implications of nanotechnology. The majority of nano news was disregarded in the prior work because of the dominance of the progress frame used when describing nanotechnology, as in many research and development and business reports. Our technique in the present study is more exhaustive than our own prior study, as well as that of Friedman and Egolf (2005), who searched for a Boolean combination of nanotechnology concepts and risk or safety concepts and also environmental or health concepts. It is comparable but somewhat broader than that of Anderson, Allan, Petersen, and Wilkinson (2005), who searched for the occurrence of a discrete list of four terms: nano, nanotechnology, grey goo, and nanobot/nanorobot. Our goal is to obtain the most complete possible set of stories about nanotechnology.

RESULTS

Our first research question is: How has the volume of print news stories focusing on nanotechnology changed over time? The impetus for this question is to understand whether news media are paying increasing attention to nanotechnologies as these enter the marketplace. Our time period of analysis begins just before the announcement of the National Nanotechnology Initiative, and runs through 2009, just over a decade. The short answer is that news coverage varies widely from quarter to quarter, with a long-term trend of growth in news attention until the mid-2000s followed by a decline. Figure 11.1 depicts this, with AP stories

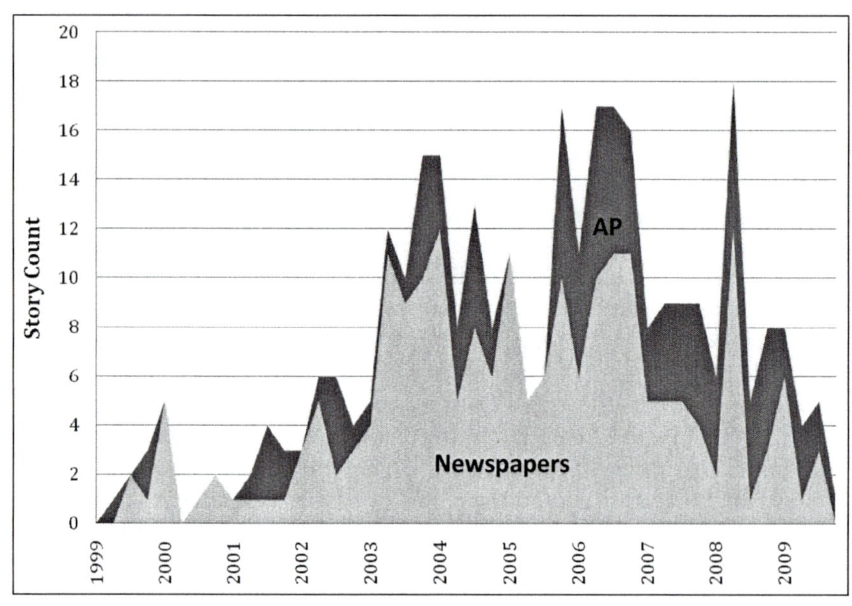

Figure 11.1 Volume of news on nano, 1999–2009.

plotted separately from the other combined sources, because it constitutes the largest single source. Across the whole period, the average number of newspaper stories per quarter is 4.7—a very modest number for such a large collection of news outlets. The average number of AP stories is 2.5. Considering the variety of top sources included in the sample, perhaps the most striking result here is the relatively low volume of news about nanotechnology, even during the most active period in 2006. Nanotechnology is not high on the news agenda and never has been.

It is interesting to compare these figures to just those stories that provide at least some attention to societal issues involving risk, potential harm or problems, the need for regulation, or other human or societal costs of developing and employing nanotechnology. About half of news stories do not cover such issues at all. Although this confirms our initial hunch that progress would dominate the news, the difference is not so great as we had expected. Given that almost half of nano news has at least some social focus, even if framed around progress, this means that science journalists are informing readers in some way about the risks as well as the benefits of new technology roughly half the time. But, again, the low overall volume of articles on nano is the key finding here.

Our second question turns more directly to the content of this flow of news: What frames do news media use for nanotechnology and how has this changed over time? We coded for the frames following the procedure described in Weaver et al. (2009). Across the whole period, the frequencies of the four frames in newspapers from all stories are as follows:

progress, 70 percent; generic risk, 17 percent; regulation, 11 percent; and conflict, 2 percent. The variation in these over time is depicted in Figure 11.2. This is a very different picture from the news primarily focused on social implications of nano: progress, 40 percent; generic risk, 37 percent; regulation 18 percent; and conflict 5 percent. In socially focused news, the progress and risk frames occur in nearly the same volume. However, in the entirety of nano news, the progress frame occurs more than three times as often as the risk frame and over two times as often as all other frames combined. This strongly supports the conclusion that nano is no exception to the rule that science journalism is dominated by a progress narrative.

One potential explanation for the dominance of the progress frame in nanotechnology news is the nature of the science itself. The term nanotechnology refers to any material or device that ranges in size from 1 to 100 nm. This definition, as it is currently used by experts and media, is intentionally broad and is not limited to a specific material or application. "Nanotechnology" encompasses both gold nanoshells, which are used to treat cancer, as well as carbon nanotubes, which are known to mimic asbestos in the lungs (Poland et al., 2008). By coding the same stories for both frame and the main application or domain, we can explore the idea that specific applications are portrayed as inherently risky and will be dominated by the risk or regulation frame, whereas other applications are portrayed as generally beneficial using the progress frame.

To answer our third research question—To what extent do newspapers frame unique nanotechnology applications differently?—we consider our nine nanotech domains: human health, medicine, consumer goods, food, energy, computers and information technology, environment, economy, and other.

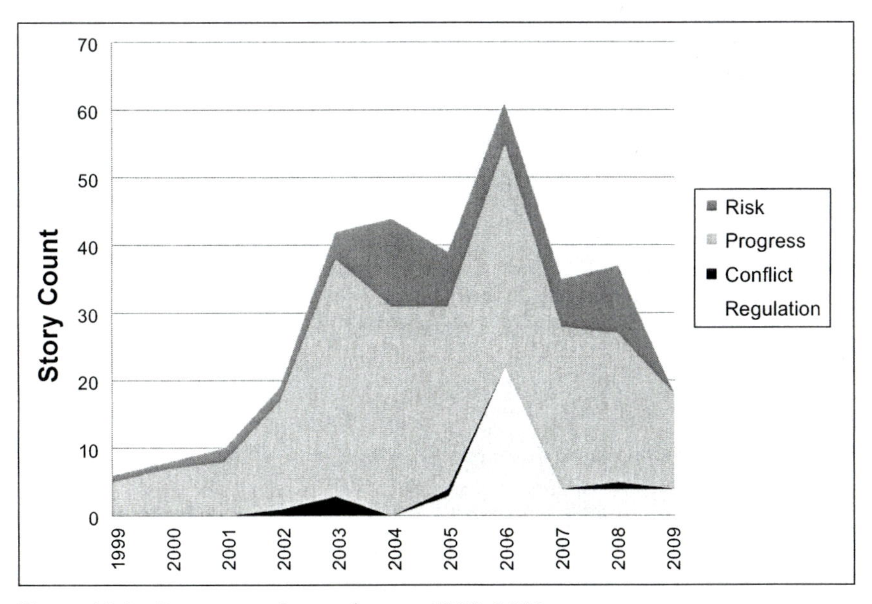

Figure 11.2 Frequency of news frames, 1999–2009.

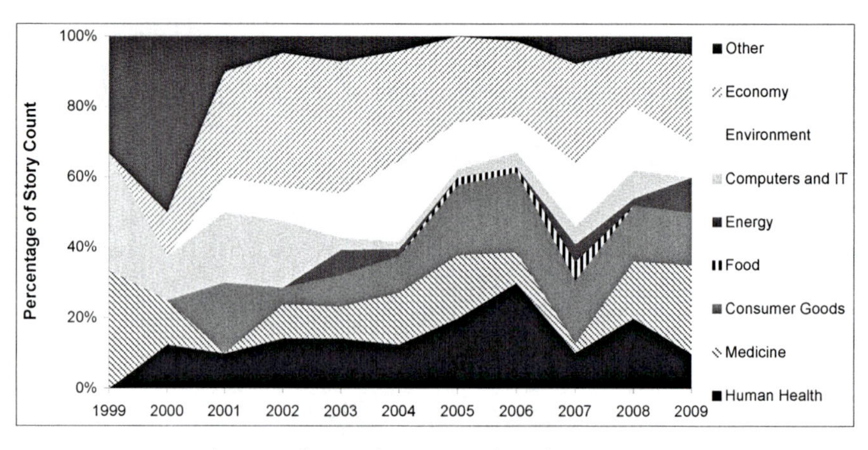

Figure 11.3 Distribution of nano domains and applications in news, 1999–2009.

The distribution of nanotechnology domains presented in the news over time shows whether media portrayal of nanotechnology's potential societal impacts has changed from 1999 to 2009. As Figure 11.3 shows, the higher percentage of "other" stories in the early years of nano coverage indicates that journalists and editors might have been unsure of the applications of nano. However, as time progressed, they seemed to find more concrete topics for storytelling, especially using applications in human health, consumer goods, environment, and economy. The domains of energy, food, and computers accounted for a relatively small percentage of their stories. The overall trend suggests that after the formative news years, the salience of topics associated with nano has remained fairly constant over time. For example, the "human health" domain has composed 10–20 percent of application coverage in nine out of the last 10 years, reaching 30 percent only one year. Even despite periodic releases of reports detailing the potential risks of nanotechnology, news coverage did not spike significantly.

When we explore the correlation between the four general frames and the individual application areas of nanotechnology, the results become more nuanced. As shown in Figure 11.4, "environment" and "human health" are dominated by the risk and regulation frames. Journalists only utilize the progress frame for nano applications in "medicine," "energy," "computers," and "economy." This is no doubt due to the expert narrative from the research establishment and corporations that nano will revolutionize these fields and begin a new industrial revolution. The potential for nano to provide environmental remediation has also been a common narrative among elites, including the NNI itself, and connection the media draws between nano and the environment seems to be a story of harm and not benefit. This division by journalists of nanotechnologies into two categories, those involving depoliticized "progress" discourse and those involving risk and regulation, has the potential to become an enduring classification in public discourse.

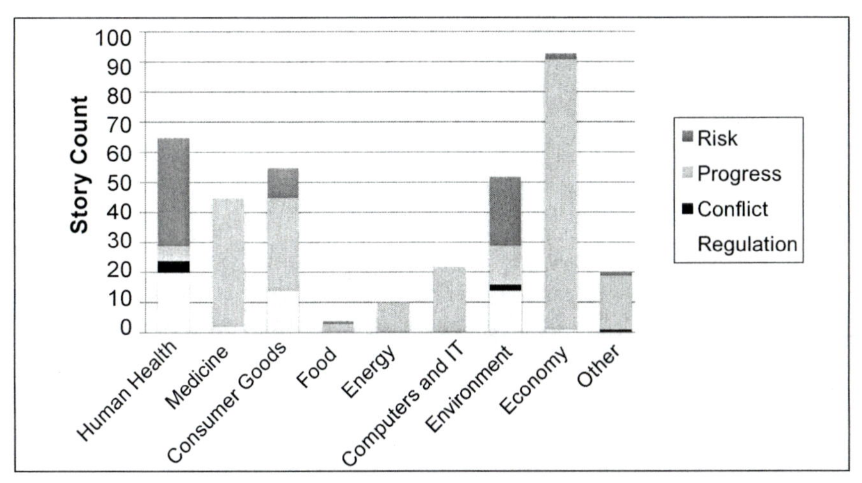

Figure 11.4 Volume of nano domains and applications in news, 1999–2009.
Note: Cumulative story count, by domain, separated by frame.

Many of the proposed advanced applications of nanotechnology have yet to come to fruition in everyday life. However, there are now available hundreds of consumer goods ranging from cosmetics to tennis rackets that use nanotechnology. Stories in the domain of "consumer goods" are mainly framed by progress (56 percent), but there are a significant number of stories framed by risk (18 percent) and regulation (26 percent). This could mean that those domains strictly portrayed in the "progress" frame could be susceptible to scrutiny by media if they become part of our everyday lives. Should current journalistic practices continue, consumer products have some indications of emerging as a contested terrain for public discourse, between the "progressive" applications such as medicine, energy, and computing, and the potentially contested applications such as environment and health.

Our final research question is directed toward assessing how news sources themselves vary their message about nano. In other words, if one were to read a single newspaper for a decade, to what extent would the result be exposure to a single dominant frame about nanotechnology? The answer is that it depends on which newspaper one were to read.

As Table 11.1 shows, some newspapers tend to stick to one frame most of the time. The *Houston Chronicle*, *San Jose Mercury News*, *Wall Street Journal*, and AP all report on nano in the progress frame at least 70 percent of the time. Other news sources such as the *Denver Post* and *New York Post* also have very high rates of progress-dominated news, though they have very few stories over all 11 years.

Other news sources are quite varied in their reporting of nano. The *Washington Post* and the *New York Times* employ a combination of risk and regulation frames in at least 50 percent of their news coverage. And despite reporting in an area with a very high concentration of nanotech companies, the *San Francisco Chronicle* devotes 46 percent of its stories

Table 11.1 Frame Variance by News Source, 1999–2009

Publication	*Story Count*	*Regulation*	*Conflict*	*Progress*	*Risk*
Denver Post	9	0%	0%	89%	11%
Houston Chronicle	51	12%	0%	73%	16%
New York Post	3	0%	0%	100%	0%
New York Times	35	9%	9%	57%	26%
San Francisco Chronicle	26	15%	0%	54%	31%
San Jose Mercury News	22	23%	5%	73%	0%
USA Today	16	6%	0%	69%	25%
Wall Street Journal	15	0%	7%	87%	7%
Washington Post	33	21%	3%	30%	45%
Associated Press	109	10%	0%	83%	7%
Total	319	12%	2%	70%	17%

to a combination of risk and regulation frames. There is a diverse message about nano coming from some news sources but not all.

In addition to frame variance, we also consider whether news sources may focus on one or a few domains or application areas of nano. We have no strong *a priori* reason to expect a pattern here. However, we note that the *Washington Post* tends to emphasize regulatory issues, and certain nanotechnology application areas have more potential for regulation in the short term, such as health and medical as well as environmental applications. The *San Francisco Chronicle* and the *San Jose Mercury News* are the hometown papers of Silicon Valley, so they might attend more closely to information technology applications. Yet no strong patterns emerge: which papers cover which topics seems at this point largely idiosyncratic.

CONCLUSIONS

These data on media coverage from 1999 to 2009 suggest several conclusions. Attention to nanotechnology in American print media is relatively low and episodic. Someone looking at news trends in 2006 would likely have concluded that nanotechnology had finally reached the news agenda and was about to take off as a public issue. Expert reports on safety and societal implications were piling up; some key advocacy groups such as Friends of the Earth and ETC Group were attracting attention with calls for regulation and a moratorium on research into certain areas; both the Environmental Protection Agency and the Food and Drug Administration were contemplating issuing regulations, and the latter did regulate nanosilver in washing machines. This emergent regulatory attention to nano attracted the attention of print news media. Under other circumstances, these activities might have continued to develop had some prominent elected officials or candidates for office devoted attention to nanotech, had advocacy groups met with a more favorable regulatory environment, or had a major risk event occurred in which nanotechnology was faulted for causing some harm or serious problem in the United States. These might have fueled a positive feedback cycle of continued growth in attention from media, more attention from the public, with more calls for policymaking, and so on (Jones & Baumgartner, 2005; Pidgeon, Kasperson, & Slovic, 2003; Tan & Weaver, 2007). None of this happened however, and following 2006 journalists turned their attention elsewhere.

Our evidence supports the general principle that unless news media have some compelling reason to do otherwise, they will tell stories of science and technology that are framed around "progress." The first four years of news in our analysis were almost exclusively progress oriented. This changed with the developments of the mid-2000s, as societal implications research was funded (including the funding for this study), and regulatory and risk frames appeared, especially for stories on environmental and human health. By 2009, news coverage was again dominated by progress frames.

Public opinion in the aggregate is often inconsistent, uninformed, or even incoherent. It is also not readily changed, certainly not in engineered directions, by the provision of more scientific information by experts or by "public information" campaigns. This can be a source of surprise to those who do not study public opinion as an object of research and it can be a source of frustration to engineers and experts in the physical or biological sciences. The latter sometimes succumb to the temptation of thinking of the public as an underachieving pupil in need of more homework or more in-depth lessons from qualified instructors. Unfortunately, from that perspective, the public does not understand itself to be in school or in need of training.

So far, news media have shown little consistent interest in nanotechnology, except as another case of the familiar story of progress. That progress story in turn has implications for the politics of technology. It tends to de-politicize what are intrinsically political processes, obscuring the actors, values, and interests at stake in technological change. Contrary to the implications of the progress frame, technology does not just happen. Public funds are spent on some research areas rather than others, and corporations invest in new product ideas as well as in favorable regulatory environments for profit-making. These are the antecedents of "progress." Nor do outcomes from technological innovation turn out to be universally positive for everyone. Some win more than others, and some simply lose. So far in the case of nanotechnology, news media in the United States have done little to make these simple facts clear.

ACKNOWLEDGMENTS

This material is based upon work supported by the National Science Foundation under Cooperative Agreement No. SES 0531184 to the Center for Nanotechnology in Society at University of California at Santa Barbara. Any opinions, findings, and conclusions or recommendations expressed in this material are those of the authors and do not necessarily reflect the views of the National Science Foundation.

NOTES

1. The National Nanotechnology Initiative (NNI) lists six major areas of "benefit and applications": everyday materials and processes (consumer goods); electronics and information technology; sustainable energy; environmental remediation; nanobiosystems, medical, and health applications; and transportation. See http://www.nano.gov/you/nanotechnology-benefits. We split the NNI's "nanobiosystems, medical and health" into two categories, human health and medicine. Human health involves potential threats to human health from nanotechnology, such as nanoparticles crossing the blood-brain barrier or exerting a toxic effect in some other way. Medicine involves applications of nanotechnology in the practice of medicine, for diagnosing or treating disease. We separated these because we want to differentiate medicine-oriented uses of nanotechnology from other uses that might have negative implications for health. Because news coverage of transportation applications is extremely infrequent, we lumped any such stories in our "Other" category, into which we also put general education stories or those dealing with military applications. We added food applications to the NNI's list because of the demonstrated salience of food-related applications to risk perception (Siegrist et al., 2007). We also added "economy" in order to reflect the fact that our own research shows that journalists sometimes describe the benefits of nanotechnology generically in terms of its potential to revitalize economies, although "economy" is not strictly speaking an application area parallel to the others. Individual nanotechnologies (e.g., carbon nanotubes) areas are not specifically identified with an application area. A news story about the similarities between carbon nanotubes and asbestos would be categorized under the "Health" application area while another story about using carbon nanotubes to shrink the size of computer processors would be categorized under the "Electronics and IT" application area.

REFERENCES

Ajzen, I., & Fishbein, M. (1980). *Understanding attitudes and predicting social behaviors*. Englewood Cliffs, NJ: Prentice-Hall.

Anderson, A., Allan, S., Petersen, A., & Wilkinson, C. (2005). The framing of nanotechnologies in the British newspaper press. *Science Communication, 27*(2), 200–220.

Berger, E. (2001, December 12). Nanotech encounters new barrier: Environmental risks rise as costs decline. *The Houston Chronicle*, p. A31. Retrieved March

2012, from http://www.chron.com/CDA/archives/archive.mpl/2001_3355884/nanotech-encounters-new-barrier-environmental-risk.html

Bijker, W. E., Hughes, T. P., & Pinch, T. (1987). *The social construction of technological systems: New directions in the sociology and history of technology.* Cambridge, MA: MIT Press.

Bridges, A. (2006, January 11). Safety laws trail behind nanotech development, report says. *San Jose Mercury News. Retrieved from Lexis-Nexus Academic.*

Brossard, D., Scheufele, D. A., Kim, E., & Lewenstein, B. V. (2008). Religiosity as a perceptual filter: Examining processes of opinion formation about nanotechnology. *Public Understanding of Science, 18*(5), 546–558.

Chong, D., & Druckman, J. D. (2007). A theory of framing and opinion formation in competitive elite environments. *Journal of Communication, 57,* 99–118.

Cobb, M. D., & Macoubrie, J. (2004). Public perceptions about nanotechnology: Risks, benefits and trust. *Journal of Nanoparticle Research, 6*(4), 395–405.

Currall, S. C., King, E. B., Lane, N. Madera, J., & Turner, S. (2006). What drives public acceptance of nanotechnology? *Nature Nanotechnology, 1,* 153–155.

Druckman, J. D. (2001). On the limits of framing effects: Who can frame? *Journal of Politics, 63*(4), 1041–1066.

Ellul, J. (1967). *The technological society.* New York: Vintage Books.

Entman, R. M. (1993). Framing: Toward clarification of a fractured paradigm. *Journal of Communication, 43*(4), 51–58.

Feder, B. J. (2002, August 19). New economy: Nanotechnology has arrived; A serious opposition is forming. *New York Times.* Retrieved March 2012, from http://www.nytimes.com/2002/08/19/business/new-economy-nanotechnology-has-arrived-a-serious-opposition-is-forming.html?pagewanted=all&src=pm

Friedman, S. M., & Egolf, B. P. (2005). Nanotechnology: Risks and the media. *IEEE Technology and Society Magazine, 24,* 5–11.

Gaskell, G., Eyck, T. T., Jackson, J., & Veltri, G. (2005). Imagining nanotechnology: Cultural support for technological innovation in Europe and the United States. *Public Understanding of Science, 14,* 81–90.

Haider-Markel, D. P., & Joslyn, M. R. (2001). Gun policy, opinion, tragedy, and blame attribution: The conditional influence of issue frames. *Journal of Politics, 2,* 520–543.

Iyengar, S. (1991). *Is anyone responsible? How television frames political issues.* Chicago: University of Chicago Press.

Jacoby, W. G. (2000). Issue framing and public opinion on government spending. *American Journal of Political Science, 44*(4), 750–767.

Jones, B. D., & Baumgartner, F. R. (2005). *The politics of attention: How government prioritizes problems.* Chicago: University of Chicago Press.

Marks, L. A., Kalaitzandonakes, N., Wilkins, L., & Zakarova, L. (2007). Mass media framing of biotechnology news. *Public Understanding of Science, 16*(2), 183–203.

Nanotechnology: Science of the small could drive big revolution. (2004, May 21). *The Houston Chronicle.* Editorial. Retrieved March 2012, from: http://www.chron.com/opinion/editorials/article/Nanotechnology-Science-of-the-small-could-drive-1982362.php

Nelson, T. E., Clawson, R. A., & Oxley, Z. M. (1997). Media framing of a civil liberties conflict and its effect on tolerance. *The American Political Science Review, 91*(3), 567–583.

Nelson, T. E., & Oxley, Z. M. (1999). Framing effects on belief importance and opinion. *Journal of Politics, 61,* 1040–1067.

Nisbet, M. C., Brossard, D., & Kroepsch, A. (2003). Framing science: The stem cell controversy in an age of press/politics. *The International Journal of Press/Politics, 8*(2), 36–70.

Nisbet, M. C., Scheufele, D. A., Shanahan, J., Moy, P., Brossard, D., & Lewenstein, B. V. (2002). Knowledge, reservations, or promise? A media effects model for public perceptions of science and technology. *Communication Research, 29*(5), 584–608.

Pidgeon, N., Harthorn, B. H., Bryant, K., & Rogers-Hayden, T. (2009). Deliberating the risks of nanotechnologies for energy and health applications in the United States and United Kingdom. *Nature Nanotechnology, 4*, 95–99.

Pidgeon, N., Kasperson, R. E., & Slovic, P. (Eds.). (2003). *The social amplification of risk*. Cambridge: Cambridge University Press.

Poland, C. A., Duffin, R., Kinloch, I., Maynard, A., Wallace, W. A. H., Seaton, A., Stone, V., Brown, S., MacNee, W., & Donaldson, K. (2008). Carbon nanotubes introduced into the abdominal cavity of mice show asbestos-like pathogenicity in a pilot study. *Nature Nanotechnology, 3*, 423–428.

Rogers, J., Shearer, C., & Harthorn, B. H. (2011). From Biotech to Nanotech: Public Debates about Technological Modification of Food. *Environment and Society, 2*, 149–169. Retrieved March 2012, from: http://dx.doi.org/10.3167/ares.2011.020109

Satterfield, T., Kandlikar, M., Beaudrie, C., Conti, J., & Harthorn, B. H. (2009). Anticipating the perceived risks of nanotechnologies. *Nature Nanotechnology, 4*, 752–758.

Scheufele, D. A., & Lewenstein, B. V. (2005). The public and nanotechnology: How citizens make sense of emerging technologies. *Journal of Nanoparticle Research, 7*(6), 659–667.

Siegrist, M., Cousin, M., Kastenholz, H., & Wiek, A. (2007). Public acceptance of nanotechnology foods and food packaging. *Appetite, 49*(2), 459–466.

Slothuus, R. (2008). More than weighting cognitive importance: A dual process model of issue framing. *Political Psychology, 29*(1), 1–28.

Slovic, P. (2000). *The Perception of Risk*. London: Earthscan.

Tan, Y., & Weaver, D. H. (2007). Agenda-setting effects among the media, the public, and Congress, 1946–2004. *Journalism and Mass Communication Quarterly, 84*, 729–744.

Weaver, D. A., & Bimber, B. (2008). Finding news stories: A comparison of searches using LexisNexis and Google News. *Journalism and Mass Communication Quarterly, 85*, 515–530.

Weaver, D. A., Lively, E., & Bimber, B. (2009). Searching for a frame: News media tell the story of technological progress, risk, and regulation. *Science Communication, 31*(2), 139–166.

Weiss, R. (2005, December 5). Nanotechnology regulation needed, critics say. *The Washington Post*. Retrieved March 2012, from http://www.washingtonpost.com/wp-dyn/content/article/2005/12/04/AR2005120400729.html

Winner, L. (1977). *Autonomous technology: Technics out of control as a theme in political thought*. Cambridge, MA: MIT Press.

Zaller, J. (1992). *The nature and origins of mass opinion*. Cambridge: Cambridge University Press.

12 Public Responses to Nanotechnology

Risks to the Social Fabric?

William R. Freudenburg and Mary B. Collins

This was the last manuscript Dr. William Freudenburg and I collaborated on prior to his passing on December 28, 2010. "Bill," as friends and colleagues called him, was unable to complete the requested manuscript revisions with me; consequently, I have made revisions to the original piece without the benefit of his scholarly perspectives and interpretations. In most cases, such revisions are entirely consistent with and integrated into our original themes, but in a few places I have inserted blocks of italicized text that represent "editorial commentary" and views that Bill may or may not have fully shared. Hopefully, the reader will not find these brief passages to be disruptive, as it has been my strong intent to leave Bill's voice and essential message of this manuscript intact.

As other chapters in this volume point out in greater detail, "nanotechnologies" are a broad and ill-defined family of products and technological innovations, united principally by an extremely small particle size. As has often been the case with earlier technologies, nanotechnology observers have spelled out a variety of views to describe the relationship between these tiny technologies and the broader public. In this chapter, we will identify four such views. The first two have been expressed primarily by non–social scientists and focus mainly on the broader public. These two views differ from one another in their prescriptions, which range from *ignoring* to *educating* the public at large. The other two views also offer differing prescriptions, but they have been developed largely by social scientists, and they devote more of their focus to the institutions that are responsible for managing new technologies.

The remainder of this chapter will be divided into three main sections. In the first, we offer a brief summary of actual survey evidence regarding the broader public's views toward nanotechnology, particularly in the United States. In the second, we spell out in greater detail the four main views just noted. As we will point out in this section, the first argument—namely, that the public should be ignored—is relatively rare, whereas the second, encouraging "public education," is considerably more widespread. The two remaining views tend to be embraced by social scientists who have thought systematically about risk and society, but they also differ considerably from one another. One of them comes from a well-known group of European social theorists working on "reflexive modernization," whereas the other comes from other social scientists—often but not exclusively from the United States—who have focused on "risks to the social fabric." In a third and concluding section, we consider the implications of these four points of view for future analyses of technological controversies more broadly, and for the sensible management of nanotechnology in particular. We end with recommendations for more empirical testing of the institutional factors that influence public perception of technological risk, with particular attention to the role played by the *trustworthiness of institutional risk managers*.

NANOTECHNOLOGY, RISK, AND PUBLICS: *WHAT DO WE KNOW?*

Later in this section we discuss what is known at the intersection of nanotechnology and risk, but to begin with, it is important to establish the most promising aspects of nanotechnology. Understanding the potential of this technology speaks to its incredible promise as well as its incredible unknowns and risks. According to the National Nanotechnology Initiative (NNI), nanotechnology is going to change the way we live, by creating new scientific applications that are smaller, faster, stronger, safer, and more reliable. These innovations fall into several main categories: *medicine*, focusing on disease diagnosis and treatment through new drug-delivery protocols that minimize harmful side effects; *energy*, focusing on developing clean and affordable renewable energy sources; *environmental quality*, focusing on affordable water purification efforts, innovative hazardous waste remediation protocols; *information and communication*, focusing on memory storage and novel computing devices; *heavy industry*, focusing on transportation and construction; and *consumer goods,* focusing on food, personal care, and sporting goods ("List of Nanotechnology Applications," n.d.).

Whereas backers of nanotechnologies tend to emphasize the impressive potential for societal benefits, other observers have emphasized potential risks of nanotechnologies, particularly in four areas—(1) *occupational safety*, as related to the manufacture or use of nanoparticles; (2) *consumer safety*, regarding physical contact with nanotechnology-based products; (3) potential *environmental damage*, resulting from manufacturing processes as well as from direct contamination of air, water, or soil by nanoproducts; and (4) potential *socioeconomic disruptions*, particularly in agriculture, raw materials, or labor. As is often the case, the intended impacts of nanotechnologies are overwhelmingly positive, whereas concerns are raised mainly by the potential for unanticipated or unforeseen consequences (see the classic assessment by Merton 1936; for a more recent example, see the statement by Marchant & Sylvester, 2006).

There is significant scientific and governmental interest in predicting public responses to the potential opportunities as well as threats of nanotechnology products, and a number of noted researchers have conducted relevant studies. In one of the more notable efforts, Satterfield, Kandlikar, Beaudrie, Conti, and Harthorn (2009) conducted a meta-analysis of existing studies, finding that about half of the public has at least some minimal familiarity with nanotechnology and that, by a ratio of 3:1, the broader public currently sees the benefits of the technology as outweighing the risks. Women, older persons, and those of lower income know less and express lower levels of confidence in the ability or willingness of government or industry to protect the public from the risk (Peter D. Hart Research Associates, 2009). A particularly striking point of this meta-analysis was that nearly half of the respondents surveyed were unsure about what nanotechnology actually is, suggesting that risk judgments are highly malleable at present (see also Peter D. Hart Research Associates, 2009; Pidgeon, Harthorn, Bryant, & Rogers-Hayden, 2009; Waldron, Spencer, & Batt, 2006) and underscoring the need for ongoing attempts to measure any changes in public perceptions of nanotechnology risks and benefits.

In addition to these general findings, many of the studies that have already been mentioned also address public perceptions specifically in the context of nanotechnological applications. For example, Pidgeon et al. (2009) found an intriguing variation between earlier observations of La Porte and Metlay (1975), when they noted that publics are increasingly "wary" toward profit-oriented applications of technology—as opposed to continuing support for "science" in general. Further, the more recent study by Satterfield et al. (2009) found that most respondents expressed reasonably high regard for "science and technology" applications in general. "Energy" applications, however, were viewed far more favorably than were applications related to "health and human advancement."

One of the things that experts who study the public and nanotechnology know quite well is the fact that there isn't just one type of *public*—rather, there are multiple types of *publics*. The work of Kahan and Rejeski (2009) (see also Kahan, Slovic, Braman, Gastil, & Cohen, 2007) has been particularly significant in disaggregating the broad category suggested by discussions about "the" public. By exposing subjects to varied forms of nanotechnology-related communication, Kahan and Rejeski found that "public attitudes are likely to be shaped by psychological dynamics associated with cultural cognition" rather than by topic familiarity (p. 87). This finding indicates not all "publics" have the same perspectives, and that if what they have found is true, then we can expect attitudes to be shaped powerfully by prevailing worldviews.

In a point that will receive additional attention later in this chapter in the nanotechnology context, a number of studies have pointed to the importance of trust and trustworthiness as related to other emergent technologies. For example, Marchant and Sylvester (2009) note what they characterized as betrayals of public trust, including views toward nuclear power after the Three Mile Island debacle (Cutter, 1984; Freudenburg, 1985); the contamination of the human food supply with genetically modified corn that had been intended only for animals; the death of a participant in a gene therapy trial; and the false assurances by the British government that "mad cow" disease would not spread to humans (for a more detailed analysis, see especially Eldridge & Reilly, 2003). Others have also found trust in government to affect public acceptance of emerging technologies, such as gene therapies (Siegrist, 2000), genetically modified food (Lang & Hallman, 2005), the acceptance of nuclear waste repositories (Freudenburg & Gramling, 1993; Pijawka & Mushkatel, 1991; Slovic, Layman, & Flynn, 1993) and proposals for offshore oil development (Freudenburg & Gramling, 1993; Picou, Marshall, & Gill, 2004).

In the nanotechnology context, a number of studies have pointed to the importance of trust and trustworthiness. In one such study that built on past work, but with a somewhat different pattern of findings, Siegrist et al. (2007) compared lay and expert perceptions of risk related to 20 potential nanotechnology applications and three non-nanotechnology applications. Unlike earlier studies that focused on controversial technologies (e.g., nuclear power, offshore oil drilling), Siegrist and colleagues found considerable similarity between experts and the broader public. Both groups assessed nanotechnology applications as being low in risk, especially when compared to risks associated with known dangers, such as asbestos exposure. Further analyses suggested that "perceived dreadfulness" of applications and "trust in government" were important factors in determining perceived risks. In another study, Macoubrie (2006) explored public concerns in relation to nanotechnology applications, finding that survey respondents had low trust in government's ability to manage risk, with more highly educated respondents expressing lower trust in government. The study also found that the participants formed perceptions based on past experiences with emerging technologies in which the limitations and dangers were initially poorly understood and later poorly managed. In yet another study, researchers in Spain studied the effect of trust of regulators when public information on new technology is scarce, specifically as related to the installation of a Centre for the Investigation of Advanced Technologies (Prades, Espluga, Real, & Sola, 2009). This facility is intended as a site for research on cleaner and more efficient use of fossil fuels, but it entails potential human and environmental health risks, because it requires the development of procedures for capturing and transporting CO_2 underground. Results revealed that, as in the case of views toward nanotechnology to date, citizens' attitudes were largely positive, but they were shaped less by technical knowledge than by citizens' tendencies to interpret an unknown technological project by placing it within an economic, social, and political context, and by relating it to prior knowledge and experiences. The findings also led the authors to hypothesize that trust and perceived risk reflected general attitudes toward a technology and the institutions responsible for overseeing it.

Policy scholars have also noticed the importance of trust in the context of nanotechnology. For example, in a prominent commentary piece, authors argued that new technology will only be successful if promotional efforts demonstrate safety. They go on to highlight case after case of *once* promising technologies that never became viable after early warnings related to safety were ignored. Although they point out that nanotechnology has made some progress toward more sustainable development in terms of safety, they continue to question its oversight (Hansen, Maynard, & Baun, 2008).

In another quite lengthy report, Europe's DEEPEN project speaks extensively about many aspects of responsible development within the nanotechnology age. Most relevant to this chapter is their discussion of related governance challenges and understanding publics. They speak to key narratives among publics that highlight uncertainty and skepticism—themes that are not as present among nanoscientists—existing alongside a climate in which both governance and regulation are developing (Davies, Macnaghten, & Kearnes, 2009). Although both of the aforementioned studies point out problematic tensions between nanotechnological development and public safety and offer some ideas about how to deal with this tension, neither one explicitly presents a theoretical foundation that might guide expectations or help manage problems before emergence.

Aside from early signs showing an emergence of work focusing on issues of trust and trustworthiness, it would be difficult to argue that work to date on public views toward nanotechnologies could be claimed to be moving toward shared agreement—with the possible exception that most studies still find that levels of public awareness are currently quite low, and that public attitudes, although modestly positive, are far from settled. Instead, as we will both argue and illustrate in the next section of this chapter, there is a need at present not just for "more research," in general, but also for more attention to the patterns of expectations that have been expressed in the existing literature. In particular, as we will spell out in the following section, we see evidence of the need for critical examination of the assumption that public reactions to nanotechnology are likely to be directly related to "objective" assessments of technological risks, or alternatively, to the level of responsibility and trustworthiness that is discernible in the actions of "responsible" risk-management institutions.

FOUR POINTS OF VIEW ON PUBLIC RESPONSE

As noted in the introductory section of this chapter, it is possible to identify four distinctive patterns in the views that have been expressed in the literature to date regarding the relationships between nanotechnology and the public. The first two focus mainly on members of the general public; the other two focus more directly on the institutions that are expected to manage the technologies and their risks. We will refer to the four patterns as (1) ignore the public, (2) educate the public, (3) reflexive modernization, and (4) risks to the social fabric; recreancy. To begin with, we discuss each perspective and then in a later section of this chapter we show how they are applicable to nanotechnology research and policy.

Predominantly Non–Social Science Points of View

Ignoring the Public

For the most part, when non–social science authors have discussed the relationships between technological risk and the broader society, they have done so with calls either for "educating" the public about "real" risk numbers, or else for removing the public from risk decisions altogether. Some of the more extreme titles decry everything from "phantom" risk, to "higher superstition," to "eco-hysterics and the technophobes" (e.g., see for Beckmann, 1973; Foster, Bernstein, & Huber, 1993; Gross & Levitt, 1994).

A few social scientists have expressed comparable views, but the voices of social scientists that have thought systematically about relationships between technology and society are clearly not the dominant ones to be expressing such views. Perhaps the most prominent exception in relatively recent work involves the sentiments expressed by the respected sociologist W. S. Bainbridge (2002):

> Modern science policy is not determined by voters in elections or referendums, nor by random samples of the population, but through informal processes in which knowledgeable opinion leaders influence decision makers in government and industry. Thus, it may be more important to study the views of knowledgeable people who want to express their considered opinions, than the views of the inarticulate and inert general public. (p. 562)

This argument offers an intriguing parallel to a much earlier finding by Goss and Kamieniecki (1983). They found that Congressional voting on energy issues generally tended to reflect the views and preferences of constituents—but that far less congruence was in evidence in cases where the Congressional representatives believed the views of their constituents to be irrational and/or ill-informed. What makes the earlier finding particularly intriguing in the context of evidence on the knowledge-deficit hypothesis, which we will summarize subsequently, is that Goss and Kamieniecki found Congressional representatives' beliefs about supposedly "irrational" constituent concerns were limited largely to public views toward nuclear power, which continued to receive high levels of federal funding even after public opposition to that technology became quite intense.

Educate the Public

The second view is considerably more widespread. Rather than arguing for an emphasis on "the views of knowledgeable people," this perspective calls for increased focus on "educating" the public more broadly. Like substantial numbers of arguments about previous cases of controversial technologies, this literature reflects what Davidson and Freudenburg (1996) termed the "knowledgeable support" hypothesis, or what Brunk (2006) has called the "knowledge-deficit" hypothesis. This second view, in other words, reflects the assumption that, if only the public were aware of the benefits of a new technology, they would accept it. Such assumptions, not surprisingly, often lead to calls for broad public education campaigns. One aim of a recent NNI project in the United States, for example, is to "identify and eradicate misunderstandings." Alongside the notion that positive evaluations of a technology will follow from being educated about it, there is often the corollary that deficits are frequently filled with information drawn from sensationalist media coverage and that this explains objections to a technology.

A reasonably representative expression of the deficit view would be the following Foresight Institute summary of a 2008 Woodrow Wilson poll:

> A recently released poll shows that the American public is largely uniformed about both nanotechnology and synthetic biology, and furthermore that the level of public awareness about nanotechnology has not changed since 2004. Perhaps there is a need for an imaginative public education effort. (AAAS EurekAlert, 2012)

At the core of this expectation is the assumption by experts that "public reluctance to accept new technologies reflects a fundamental failure to understand what scientific and other expertise has established as true" (Brunk, 2006, p. 178). Ascribing a privileged status to expert knowledge, this view strongly recommends "public education" or public relations campaigns to overcome "uninformed" or irrational perspectives, and to persuade the public to accept and adopt the views of scientific authorities and public policymakers. Although this viewpoint has been frequently questioned over the last several decades, it remains an enduring conviction in the minds of many proponents of new technologies—perhaps in part because many of them are convinced that their own enthusiasm is based "strictly" on their knowledge of the available evidence, rather than also being based on their values, and on their choices about which pieces of evidence deserve more or less emphasis.

Predominately Social Science Points of View

With very few exceptions, however, social scientists' treatments of risk-and-society issues have expressed starkly different views from the two just outlined—although there are at least two sharply differing perspectives within the social science literature as well. The roots of the social science work date back at least to the time of Habermas (1970) and his questions about what he termed a "legitimation crisis," but much of the early work on this "crisis" tended to emphasize *economic* risks and efforts to maximize economic performance (e.g., see Block, 1987; Habermas, 1973; O'Connor, 1973; Offe, 1987). Particularly beginning in the 1980s, however, increasing numbers of sociologists, in particular, began to focus on potential challenges to legitimacy relating to *technological* risks—and as the amount of attention grew, so did the degree of divergence between two key bodies of social science work.

Today, even if the focus is limited merely to those who have offered macrosociological lines of analysis, there are at least two main patterns. The first involves well-known European social theorists of "reflexive modernization," whereas the second involves the work of several scholars, predominantly but not exclusively from the United States, whose work emphasizes instead what Short (1984) first termed "risks to the social fabric."

Reflexive Modernization

One of the most vital bodies of sociological work in recent decades has involved the scholarship on "reflexive modernization," two of the most important proponents of which have been Ulrich Beck and Anthony Giddens. Although both of them place a good deal of emphasis on large issues of risk and "reflexivity"—roughly, the act of reflecting on, referring to, and hence potentially affecting the choices being made—the two authors develop their arguments in ways that also include a number of important differences, particularly with respect to the views they express regarding the roles of scientific and technological experts.

In his *Modernity and Self-Identity*, for example, Giddens (1991) depicts the breakup of traditional communities, in conjunction with globalization processes, as freeing individuals to reflect on their actions and to develop or choose their identities. He sees scientific knowledge as playing an increasingly important role in that process, providing key potential inputs for the exercise of "dialogic democracy"—a concept involving open communicative exchanges, independent from formal political institutions, and carrying the spread of social reflexivity in ways that condition both everyday life and collective action (p. 115). Beck's *Risk Society* (1992), alternatively, could scarcely offer a less similar view of the role of science. In his view, the legitimacy of scientific and technological rationality has been called into question by the inability to control the very risks that scientific expertise has brought into being—risks that have become global in scope. As he puts it, "One can possess wealth, but one can only be afflicted by risks" (p. 23). Beck, accordingly, argues that "'the sciences' monopoly on rationality is broken" (p. 29), with citizens finding solidarity not through a shared appreciation for scientific expertise, but through a shared skepticism toward it, inspired in good measure through their shared exposure to risks (see also Picou & Gill, 1999).

These two European theorists do tend to share comparable views, however, in the extent to which they emphasize what Giddens (1990) calls "high-consequence risks." In the work of Giddens, this phrase refers to truly formidable, global-scale risks, ranging from "nuclear warfare and nuclear winter to chemical pollution of the seas" (pp. 124–125)—the types of risks that he characterizes as creating a society-wide concern, so pervasive that it "transcends all values and all exclusionary divisions of power" (p. 154). For this reason, this issue is sometimes described as involving "transcendence." Beck, similarly, pays special attention to nuclear, chemical, ecological, and genetic engineering risks, which he sees as involving "uncontrollable consequences" that are not limited in time or space (Beck, 1995, p. 31), and which he sees as presenting risks so massive as to "destroy the . . . principal pillars of insurance," being too large to be underwritten even by modern-day insurance companies (p. 127).

RISK AND "THE SOCIAL FABRIC"

By contrast, a growing number of other social scientists, based principally but not exclusively in the United States, have offered analyses that differ quite starkly from the views of Giddens and Beck, and few of them would describe themselves as working from within a perspective of "reflexive modernization." Perhaps partly for that reason, a common tendency is to contrast the "European" and "U.S." approaches (see also Cohen, 2000). In substantive terms, however, although this U.S.-based work does tend to differ from the work of both Giddens and Beck on the issue of transcendence, as we will spell out subsequently, the same scholars' work on the key issue identified in the opening pages of this chapter, involving the roles and trustworthiness of scientific and technical experts in democratic systems, involves not so much a "counterpoint to" as a position that is essentially *between* the views of the two European theorists. We will explore both of these points—transcendence and the roles of scientific/technical experts—in turn.

We deal first with the issue of transcendence. Like Giddens and Beck, U.S.-based risk researchers have devoted a good deal of attention to nuclear, chemical, and ecological risks—the work of Perrow (1984), for example, originally grew out of investigations for the President's Commission on the Accident at the Three Mile Island nuclear facility in 1979—but much of the work has treated "the socially salient risks" as being "precisely those risks that do *not* 'transcend all values and all exclusionary divisions of power'" (Freudenburg, 2000, p. 112).

The distinction is particularly clear because, in many respects, the early U.S. work on technological risks and disasters grew out of an earlier emphasis on so-called "natural" hazards and disasters, such as earthquakes and tornadoes. The earliest of the well-known studies of what have sometimes come to be called "technological disasters," in fact, was Erikson's (1976) analysis of a flood in West Virginia, which was caused when a coal mine's earthen dam gave way, leading to the flooding of a long mountain valley. That flood did indeed "transcend" most social divisions at least within that valley, leaving over 100 people dead and over 1,000 homeless, but as Erikson pointed out in considerable detail, rather than *creating* a community—whether "of anxiety" or of any other variety—the experience of risks from this human-caused disaster led to what the subtitle of his book called "the *destruction* of community." Erikson emphasized this point, in fact, partly because the phenomenon was so different from the pattern that had more commonly been seen in so-called "natural" hazards and disasters up to that time.

As had already become well known within the disaster literature by the 1970s, in other words, natural disasters did indeed tend to be characterized by what that literature had come to call a "therapeutic community" (e.g., see Barton, 1969). In the commonly observed pattern, citizens from all walks of life would come together, more or less spontaneously, to offer aid to the victims in the aftermath of a natural disaster. As was noted in a later review of the differences between natural versus "anthropogenic" or technological disasters, however, such reports have remained almost completely absent from cases where the risk or disaster has been of human origin, where instead the common pattern involves what that later review characterized as "corrosive communities" (Freudenburg & Jones, 1991; Kroll-Smith & Couch, 1993); (see also Erikson, 1976, 1995; Picou, Gill, Dyer, & Curry, 1992).

The second pattern of difference involves issues of trustworthiness and the roles of experts. On this issue, as noted already, the U.S.-based literature presents not so much an opposing view to the arguments put forth by Beck and Giddens, but instead, a view that is intermediate between those of the two European theorists. In the U.S.-based literature on technological risk, the central tendency is to see *most* technological systems as having worked properly, the clear majority of the time—but with the added point that even occasional exceptions can be profoundly troubling, leading to the creation of what Short (1984) has termed "risks to the social fabric." In what may be the most explicit statement of this perspective, Freudenburg (1993) traces the reasons back to European social theoretical frameworks of an earlier vintage, deriving largely from Durkheim (1893/1933) and Weber (1919/1946).

Much as Durkheim spelled out, Freudenburg (1993) argues, the division of labor in society does appear to have permitted tremendous increases in the overall level of expertise and prosperity enjoyed by present-day citizens of the industrialized world—but it has done so with one important catch. When Durkheim first called attention to the division of labor, he referred approvingly to what he called "organic solidarity," seeing the coordination of differing specializations as being relatively unproblematic. With increased specialization, Durkheim argued, different kinds of people would come to need each other just as much as do different organs of the body, with the heart and the stomach, for example, each filling its own specialized role. Unlike stomachs, however, humans have the capacity to discern specialized interests that can differ significantly from the needs or interests of the collectivity.

Although Durkheim did not treat such possibilities as being problematic, they lie at the core of what Freudenburg (1993) calls "recreancy"—"the failure of institutional actors to carry out their responsibilities with the degree of vigor necessary to merit the societal trust they enjoy" (p. 909). In related work, Siegrist, Earle, and Gutscher (2003) liken "recreancy" to "confidence," which they argue to be unlike "trust" in being affectively neutral. The ability to discern special or non-shared interests also appears to be important in terms of what Short (1984) has called "risks to the social fabric."

The reasons were initially suggested long ago in Weber's (1919/1946) discussion of what it meant to live in a world of "intellectualized rationality." What made the world a "rational" one, in Weber's view, was not that the denizen of modernity could be expected to know more about the world around us, but very nearly the opposite.

> Unless he is a physicist, one who rides on the streetcar has no idea how the car happened to get into motion. And he does not need to know. He is satisfied that he may 'count' on the behavior of the streetcar. . . but he knows nothing about what it takes to produce such a car so that it can move. The savage knows incomparably more about his tools. (pp. 138–139)

Far more than was the case for our great-great-grandparents, however, the citizens of today's world tend to be not so much in control of as dependent on our technology. We need to "count on" that technology to work properly—not just in principle, but also in practice. As a result, we are dependent not just on the technologies, but also on the social relations that bring them into being, involving whole armies of specialists, most of whom have areas of expertise that we may not be competent to judge, and many of whom we will never even meet, let alone have the ability to control. This perspective, accordingly, is neither as pessimistic as the expectations of Beck (according to whose arguments, as will be recalled, technology is essentially out of control), nor as optimistic as expected by Giddens (according to whose arguments the inputs of science and technology are relatively unproblematic). Instead, the U.S. work suggests that the vast majority of present-day technology can be "counted on" to work properly, the vast majority of the time—but that citizens may be acting quite rationally if they become concerned when some key element of the sociotechnical system sends a "signal" (cf. Slovic, 1987) that matters are not being controlled as safely as ought to be the case.

IMPLICATIONS FOR RESEARCH AND NANOTECHNOLOGY POLICY

As noted at the outset, public views toward nanotechnology are currently low in salience and information levels, meaning that few predictions about future trends can currently be made with confidence. In general, attitudes tend to become more predictable as they become more deeply embedded into a person's established ways of thinking about and making sense of the world (see also Heberlein, 1981). Some technologies reach that level of salience, others do not—and still others achieve a more mixed status. The work of Kahan and his colleagues suggests that future public views toward nanotechnology within the U.S. public may well depend significantly on the worldviews of individual citizens, which are by nature "mixed" or differentiated—and experience with other technologies can show us examples of different views emerging in different cultures or nations. In many of the nations of Europe, for example, views toward genetically modified organisms became actively and intensely hostile during the latter years of the twentieth century, whereas little such opposition has materialized in the United States, at least to date.

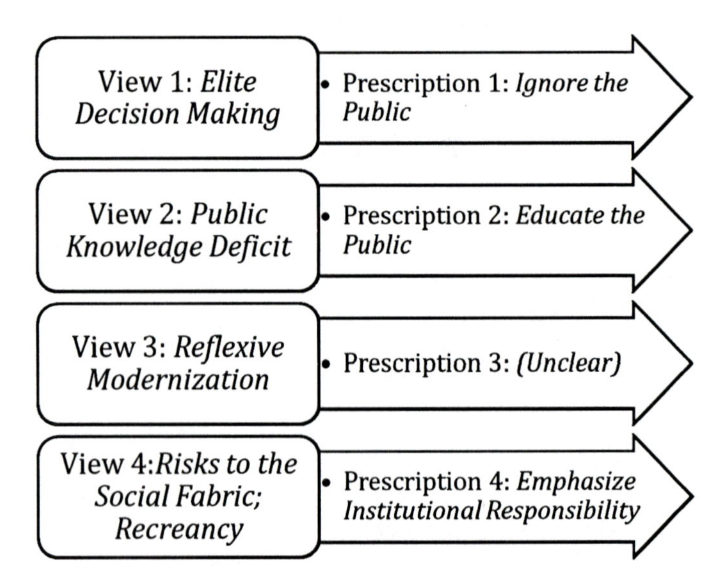

Figure 12.1 Four main perspectives and prescriptions on public views toward nanotechnology.

Clearly, the only way to know with certainty what will be seen with future views toward nanotechnology will be for ongoing research to be continued. At the same time, it is not too soon to begin thinking more systematically about the four predominant views that have been identified in this chapter, and to consider the implications that can be derived from past experiences with other forms of technology. As one way to begin that process, we offer a simplified typology, which is represented in Figure 12.1. As suggested, the four perspectives all contain at least implicit conceptual model views—and at least three of the four also point to policy prescriptions that may be relevant in thinking about nanotechnology. The policy prescription, or lack thereof, for the third view, "reflexive modernization," will be discussed in great detail later in this section.

The first perspective—described previously in terms of its implicit policy prescription, namely, to "ignore the public"—reflects at least an implicit adherence to a model of elite decision making. In pragmatic terms, perhaps the most useful way to think about this model (and its associated prescription) is provided by Stone's (1980) distinction between "democratic power" and "systemic power." As Stone notes, organized business interests do not win all battles in American politics, yet they do tend to win more often than they lose, particularly over the long run. Their influence, in his terms, reflects "systemic" power—a complex concept, but one that can usefully be thought of in terms of the influence that business interests can have through factors that go well beyond mere numbers, ranging from campaign contributions and a reputation for political power through the sheer fact that they "know how to get things done"—in part because at least the leaders of such organizations can make decisions without going through a complex or lengthy process of negotiating with hundreds or thousands of their fellow citizens. "Democratic" power, by contrast, reflects the power of the ballot, plus the fact that business interests can be vastly outnumbered by their opponents—at least at times when the political salience of an issue is high. Stone's perspective predicts that, so long as the future development of nanotechnology offers few or no nasty surprises of the sort that would send a Slovic-style "signal" to the broader public, then backers of the technology might find it reasonable to expect a future of benign neglect from the public at large.

As a matter of policy, an explicit endorsement of "elite decision making" is rare, but this approach remains common in practice—more akin to a de facto adoption of this strategy. For example, some have pointed out that the public has little role in decisions about the proliferating commercial uses of nanotechnology—applications that are often developed under a cloak of secrecy by industries seeking to protect their proprietary interests (Suppan, 2011). In discussing nanotechnology applications in agriculture, Suppan suggests,

> Efforts to determine appropriate safety regulations should begin with a framework of comparative technology assessment to assess whether a nanotechnology application is the optimal means for achieving a

> specific technological or public interest goal. It appears that investors hope that products will become ubiquitous and "accepted" before risk research reveals harms so prevalent and/or severe as to force withdrawal of the product from commerce. (p. 4)

In direct opposition to the "ignore the public" approach, some very interesting public engagement initiatives have taken place in the UK and United States from which much can be learned. Doubleday (2007) summarized the findings from six major initiatives:

> The six projects reported strikingly similar responses to [a] broad set of questions about nanotechnology and the social and policy issues it raises. Public attitudes showed common patterns in both the USA and UK, and similar issues were raised both by the citizens' consensus reports and by the research reports written by analysts. The four common themes that emerged were first an expectation that nanotechnology will deliver benefits; second, a concern about the management of unforeseen risks; third, that innovation may not be directed towards the kinds of social goals that the public engagement processes identify; and finally that science and technology policy should be more open to public involvement. (p. 217)

However, mechanisms of public participation, although crucial, do not come without challenges. Rogers-Hayden and Pidgeon (2008) point out that in bringing the public in at an early phase, before significant research and development, it may well be difficult for the public to "conceptualize" (p. 1012) nanotechnology because they know little about it or about the products that utilize nanomaterials. This then may lead to conflicting views and visions between the public, policymakers and experts, and may make it hard for the relevant institutions to incorporate public views. As more is learned about effective ways to institutionalize public involvement, there will be more meaningful movement away from an implicit "ignore the public" approach to policy.

The second perspective, by contrast—characterized by frequent calls to "educate the public"—often reflects the belief that it would be a risky bet to hope for simply benign neglect in the absence of the Slovic-style signal. Indeed, it is not difficult to think of other technologies that were once seen as tremendously promising but that were stopped or dramatically slowed by public opposition—nuclear power, offshore oil development, genetically modified foods in Europe, even fluoridation of water in a number of communities. The underlying concerns that are expressed in this literature, in other words, may be utterly reasonable. The key question has to do with the assumption that opposition is likely to reflect a "public knowledge deficit" that could somehow be corrected by providing the public with more or better information. In our judgment—which appears to be widely shared by those who have studied such "public education" efforts in the past—that expectation is fundamentally flawed.

Clearly, "information" can have some effect; the question is, what kind of effect? In what surprisingly remains one of the few studies to have examined the question explicitly, Reed and Wilkes (1981) found knowledge to have an interaction effect with confidence and political ideology—generally along the lines that would later be spelled out in much greater detail by Kahan et al. (2007): Among residents in the Reed-Wilkes Massachusetts statewide study who had greater confidence in environmentalists or were politically liberal, in other words, increased knowledge was associated with greater opposition to nuclear power; it was only among those who lacked confidence in environmentalists or were politically conservative that the study found the "expected" result of information leading to greater support.

That same year, a classic analysis by Mazur (1981) predicted that "unexpected" results should actually be expected to be the more common ones—and a dozen years later, Flynn, Slovic, and Mertz (1993) confirmed that prediction, documenting the effects of an expensive "public education" initiative that resulted in a "risk communication fiasco." At the urging of a well-connected public relations firm, friends of the nuclear industry spent a sizable sum on efforts to "educate" the Nevada public about the benefits of hosting a nuclear waste repository—and after the well-coordinated public relations "initiative," statewide opposition to the repository showed a significant increase, rather than a decrease.

In a more systematic assessment, Davidson and Freudenburg (1996) reviewed literally all the studies they could find that provided actual evidence on what they characterized as one of the hypotheses that had been most often stated and tested in the empirical literature, namely, the argument that women had higher levels of concern about nuclear power due to a lower level of knowledge. As the summary of their analysis noted, the hypothesis consistently failed to be supported. Although women do tend to have lower levels of technical knowledge than men, 11 of the 16 relevant studies either fail to support the expectation that knowledge will be correlated with lack of environmental concern, or else actively contradict this expectation. The evidence, moreover, continues to indicate that the simplest plausible explanation for the discrepant findings is to be found in the methods of the relevant studies. (cf. Reed & Wilkes, 1981; see also Lopes, 1991).

Given the obvious possibility that supporters of embattled technologies might generally be tempted to distract attention away from the potential problems of their technologies by attempting to call into question the credibility or legitimacy of those who oppose them (cf. Freudenburg & Gramling, 1994; Freudenburg & Pastor, 1992), it is almost surprising that so little attention, to date, has been devoted to the possibility that such characterizations of opponents could be motivated by political and economic interests, as well as or instead of scientific interests. Although arguments about the asserted "ignorance" or "irrationality" of critics may continue to be useful as political ammunition for facility or technology supporters, these arguments appear to have little empirical or scientific merit (Davidson & Freudenburg, 1996).

As related to nanotechnology, those advocating an "educate the public" approach have been said to hold a "pave the way" viewpoint in seeking to encourage public acceptance (Sylvester, Abbott, & Marchant, 2009). Those subscribing to this viewpoint believe (based on lessons learned from unsuccessful introductions of other emergent technologies), that a government that sits on the sidelines instead of taking an active role in educating and guiding public opinion risks public rejection of the new technology in question. Advocates of this viewpoint (see Piegorsch & Schuler, 2008; Schuler, 2004) feel that governments should take decisive steps to educate the public about nanotechnology benefits and risks, thus making way for acceptance. As Sylvester et al. go on to point out, these scholars are not advocating that the public be manipulated or misled, but they clearly feel that the public is vulnerable to misinformation. As Schuler sees it, the government needs to be proactive in acting quickly to educate the public to assure that some misguided perception of risk does not cloud the "real" and less threatening risk of nanotechnology, thus jeopardizing its acceptance.

We concur with others who contend that it is misguided to believe that government either can or should play the role of "properly" educating the public (see also Sylvester et al., 2009). Regardless of what the government would choose to emphasize—the balance of risk versus benefit or a path of aggressive development versus precaution—the various groups within society will hear messages differently. Such messages will interact with other information, such as that put forth by the news media and other forms of influential media, with uncertain results. Although the influence of the media on public opinion is still debated (see also Petersen, Anderson, Wilkinson, & Allan, 2007; Wilkinson, Allan, Anderson, & Petersen, 2007), it seems clear that public opinion is not a unitary, malleable force that can be readily directed by a paternalistic hand of government.

The third perspective—the "reflexive modernization" perspective most often associated with Giddens and Beck—is one that has received a good deal of attention and respect among social philosophers, as well as in social science circles. At the same time, however, it appears to offer few if any prescriptions that would seem relevant at present to nanotechnologies. At least in our view, and based on our understanding of the research that has been done to date by our colleagues in a range of disciplines, there appears to be little reason to expect that nanotechnologies will present the kinds of formidable, global-scale risks emphasized by these theorists, ranging from "nuclear warfare and nuclear winter to chemical pollution of the seas" (Giddens, 1990, p. 125). It is always possible that such ominous implications will emerge in the future, but until or unless that happens, we see the work of Giddens, Beck, and their associates as having little relevance to future research, thinking, *or* policy decisions regarding nanotechnologies.

Some might argue that reflexive modernization would advocate the "precautionary principal" as the proper prescription in nanotechnology development, as heightened social concern would lead to greater risk aversion and need for caution. As adherents to reflexive modernization assert, technological development has created a situation of uncertainty, or in the words of Beck (2003),

> [T]hrough our past decisions about atomic energy and our present decisions about the use of genetic technology, human genetics, nanotechnology, and computer science, we unleash unforeseeable, uncontrollable, indeed even incommunicable consequences that threaten life on earth.

In other words, the notion of "incommunicable consequences" implies that society does not have the intellectual capacity to appreciate or understand the risks before it; constant anxiety of the unknown has become a defining feature of modern life (Furedi, 2009). In response to this, the reflexive modernization perspective could be seen as advocating the precautionary principal, "worst case thinking" (Clarke, 2006), or the idea that "science must not run ahead of public opinion" (Furedi, 2009, p. 203). The reflexive modernization viewpoint might therefore advocate that steps be taken to protect human health and the environment even if there is no clear evidence of nanotechnology-based harm. This goes well beyond a "wait and see" approach. In the words of the European Environment Agency, "forestalling disasters usually requires acting before there is strong proof of harm, particularly if the harm may be delayed and irreversible" (p. 13).

The outcome we see as more plausible—the fourth perspective—is that other, less spectacular problems may emerge, namely, the kinds of problems raising questions about recreancy and about "risks to the social fabric." If future developments in nanotechnology were indeed to lead to cases where even a small number of actors show such intense interest in short-term profits that they were to create significant threats to public health or the environment—hence, in Slovic's terms, sending to the broader public a "signal" that the risks are not in fact being managed as responsibly as all of us would hope—that could indeed turn into the kind of opinion-swaying event that would lead to "the wrong kind" of public education, leading the public to express intense and politically effective opposition, not just to that particular company, but to the nanotechnology industry more broadly. In an illustration of this, the widely publicized "Magic Nano" incident, in which approximately 100 people became ill after using a particular cleaning product, led to an international media storm and calls for a moratorium on *all* nanotechnology products—until it was learned that in fact the product contained no nanoparticles (Abbott, Sylvester, & Marchant, 2009). In discussing this incident, the same authors indicated that

> we have little doubt that the public reaction to a major 'incident' involving nanotechnology would be far worse if in addition to any concrete harm the technology may have caused, it also appeared that governments had sat idly by and allowed that harm to happen. (p. 4)

Issues of recreancy and "risks to the social fabric" are echoed by Sylvester et al. (2009), who describes those concerned with "upstream management" of nanotechnology as being more concerned with harming democracy, liberalism, and trust in government rather than with the damage to the progression of nanotechnology development specifically. He describes this group of policymakers and scholars as seeing public opinion as being intrinsically deserving of respect and as advocating public involvement from the outset in decisions on research, development and future applications (see also Burri & Bellucci, 2008). Sylvester et al. note that the opinions of this upstream group value public opinion in governance processes even if it leads to the rejection of beneficial technology in cases where it is in violation of core public values or when it is seen as being overly influenced by industry.

If the "prescription" of this fourth perspective is to emphasize institutional responsibility, there is some encouraging evidence that influential bodies are also thinking along these same lines in moving from the deficit model ("educate the public") to that of encouraging meaningful upstream public participation as one way of building trust. For example, the 2004 report by the UK's Royal Society and Royal Academy of Engineering (RS/RAE report) acknowledges the importance of public inclusion in the early stages of technology development (see also Petersen et al., 2007), particularly in terms of values and to encourage greater institutional trust. As Petersen et al. point out, a central aim of creating a true "dialogue" with the public is to encourage a meaningful debate early on, before opposing positions become entrenched with their constituencies. Although the UK has taken some substantive steps toward public engagement, there needs to be more in the way of developing practical mechanisms and strategies for engagement. Nevertheless, the RS/RAE report represented a seminal shift toward greater acceptance of views emphasizing democratization in the management of emerging technology. This is largely consistent with our focus on the need to emphasize institutional responsibility in the management of nanotechnology and the importance of public engagement as one potential vehicle to foster much needed public trust.

The final implication we derive, accordingly, is one that often comes as a surprise to our colleagues in other disciplines who are currently working hard to realize the tremendous potential for benefits from the future development of nanotechnologies. At least in private conversations, the most common worries expressed by those colleagues have to do with the concern that almost any sign of skepticism might lead the public to develop a nearly hysterical reaction against nanotechnologies, leading in turn to the premature abandonment of richly promising possibilities for the future. Our conclusion, in contrast, is that those who are most optimistic about the potential for nanotechnologies may also be the ones who ought to work the hardest to maintain a perspective of careful, scientific skepticism. If potential problems can be identified in advance—which is to say, before millions of dollars have been invested in manufacturing and distributing faulty or risky products—few members of the broader public are likely to notice or care. That, after all, would just be evidence that all of us are doing our jobs, and doing them in responsible ways. The way to get the public's attention—the worst possible way to do so—would be by looking the other way, or by failing to be adequately vigilant about risks to public health or the environment that could have been stopped, if only the responsible scientists and technologists had indeed been doing their jobs responsibly.

CONCLUDING REMARKS

Nanotechnology offers unique opportunities to study the factors that influence public acceptance of new technologies. With the current public knowledge of nanotechnology being low, but attitudes still being basically favorable, social scientists engaged in the study of risk perception have an important opportunity to study the development of public perceptions—doing so in a context that is still largely free of documented harms, deeply embedded political and economic influences, and powerful pro- or anti-nanotechnology social movements.

As we have discussed, "public education" campaigns can vary in the extent to which they achieve the usual goal of creating favorable impressions of a technology. In our view, at least part of the reason is that they miss an essential aspect of what it means to live in an advanced technological society. Today's technological specialization effectively requires a reliance on specialized experts and institutions, and it is important to keep this fact in mind when assessing risk tolerance and the public proclivity to trust.

Researchers face no shortage of questions needing to be asked. A number of those questions do focus on the public: What information does the public need? How should such information be conveyed? How does this information change depending on individual and group characteristics? Do the answers differ in the absence of experience-based risk information and the establishment of technology-specific governance entities? Other questions focus on the relationships between the public and the institutions bearing responsibility for developing the new technologies, or on the institutions themselves: What factors will influence the public acceptance of new institutions charged with risk management of nanotechnology? How do perceptions of institutional trustworthiness affect public views toward a new technology? How can society promote the development of greater institutional responsibility in technology regulation and risk management? Would an official policy of "precaution" affect public perceptions of risk governance institutions?

Perhaps the greatest need, however, is to move beyond the assumption that members of the public simply develop their views in response to the information provided by scientists and public officials, while somehow ignoring the "information" they receive about the performance of institutions that are entrusted with regulating and managing the technologies in question. At a bare minimum, that widespread assumption needs to be supplemented with a complementary approach—an institution-based perspective that sees public responses as embedded in the larger societal contexts of a technologically advanced society, and one that views institutional performance and trustworthiness as playing important roles in the public reactions that are likely to emerge.

ACKNOWLEDGMENTS

This material is based upon work supported by the National Science Foundation and Environmental Protection Agency under Cooperative Agreement No. DBI 0830117 to the UC Center for Environmental Implications of Nanotechnology at UCLA and University of California at Santa Barbara. Any opinions, findings, and conclusions or recommendations expressed in this material are those of the authors and do not necessarily reflect the views of the NSF or EPA. This material has not been subjected to EPA review and no official endorsement should be inferred.

I would also like to acknowledge Barbara Herr Harthorn—who has provided me with immeasurable support as I continue my education without Bill—thank you for your kindness and your advice; Nick Pidgeon who reviewed this article and provided extremely useful feedback; and, finally, Bill Freudenburg; who originally captivated me with his writing, then became my mentor, and gave me his time and intellectual support with great generosity.

REFERENCES

AAAS EurekAlert. (n.d.). In *Foresight Institute*. Retrieved March 14, 2011, from http://www.foresight.org/nanodot/?p=2858

Abbott, K. W., Sylvester, D. J., & Marchant, G. E. (2009, June 23). Transnational regulation of nanotechnology: Reality or romanticism. *International Handbook on Regulating Nanotechnologies*. Retrieved December 2, 2011, from http://ssrn.com/abstract=1424697

Bainbridge, W. S. (2002). Public attitudes toward nanotechnology. *Journal of Nanoparticle Research, 4*, 561–570.

Barton, A. H. (1969). *Communities in disaster: A sociological analysis of collective stress situations*. Garden City, NY: Doubleday.

Beck, U. (1992). *Risk society: Towards a new modernity*. London: Sage.

Beck, U. (1995). *Ecological enlightenment: Essays on the politics of the risk society*. Atlantic Highlands: Humanities.

Beck, U. (2003). The silence of words: On terror and war. *Security Dialogue, 34*(3), 255–267. doi:10.1177/09670106030343002

Beckmann, P. (1973). *Eco-hysterics and the technophobes*. Boulder, CO: Golem Press.

Block, F. (1987). *Revising state theory: Essays in politics and postindustrialism.* Philadelphia, PA: Temple University Press.
Brunk, C. G. (2006). Public knowledge, public trust: Understanding the "knowledge deficit." *Community Genetics, 9*(3), 178–183.
Burri, R. V., & Bellucci, S. (2008). Public perception of nanotechnology. *Journal of Nanoparticle Research, 10,* 387–391.
Clarke, L. (2006). *Worst cases: Terror and catastrophe in the popular imagination.* Chicago: University of Chicago Press.
Cohen, M. J. (2000). Environmental sociology, social theory, and risk: An introductory discussion. In M J Cohen (Ed.) *Risk in the modern age: Social theory, science, and environmental decision-making*, pp. 3–34. New York: Palgrave.
Cutter, S. L. (1984). Risk cognition and the public: The case of Three Mile Island. *Environmental Management, 8*(1), 15–20. doi:10.1007/BF01867869
Davidson, D. J., & Freudenburg, W. R. (1996). Gender and environmental risk concerns: A review and analysis of available research. *Environment and Behavior, 28*(3), 302–339.
Davies, S., Macnaghten, P., & Kearnes, M., (Eds.). (2009). *Reconfiguring responsibility: Lessons for public policy* (Part 1 of the report on Deepening Debate on Nanotechnology). Durham, UK: Durham University.
Doubleday, R. (2007). Risk, public engagement and reflexivity: Alternative framings of the public dimensions of nanotechnology. *Health, Risk & Society, 9*(2), 211–227. doi:10.1080/13698570701306930
Durkheim, E. (1933). *The division of labor in society.* New York: The Free Press. (Original work 1893)
Eldridge, J., & Reilly, J. (2003). Risk and relativity: BSE and the British media. In N. Pidgeon, R. E. Kasperson, & P. Slovic (Eds.), *The social amplification of risk*, pp. 138–155. New York: Cambridge University Press.
Erikson, K. T. (1976). *Everything in its path: Destruction of community in the Buffalo Creek flood.* New York: Simon & Schuster.
Erikson, K. T. (1995). *A New species of trouble: The human experience of modern disasters.* New York: W. W. Norton & Company.
European Environment Agency. (2001). *Late lessons from early warnings: The precautionary principle 1986–2000.* Copenhagen: EEA. Retrieved December 2, 2011, from http://www.eea.europa.eu/publications/environmental_issue_report_2001_22
Flynn, J., Slovic, P., & Mertz, C. K. (1993). The Nevada initiative: A risk communication fiasco. *Risk Analysis, 13*(5), 497–502. doi:10.1111/j.1539-6924.1993.tb00007.x
Foster, K. R., Bernstein, D. E., & Huber, P. W. (Eds.). (1993). *Phantom risk: Scientific inference and the law.* Cambridge, MA: MIT Press.
Freudenburg, W. R. (1985). Nuclear reactions: Public attitudes and policies toward nuclear power. *Policy Studies Review, 5*(1), 96–110.
Freudenburg, W. R. (1993). Risk and recreancy: Weber, the division of labor, and the rationality of risk perceptions. *Social Forces, 71*(4), 909–932.
Freudenburg, W. R. (2000). The "risk society" reconsidered: Recreancy, the division of labor, and risks to the social fabric. In M. J. Cohen (Ed.), *Risk in the modern age*, pp.107–122. New York: Palgrave.
Freudenburg, W. R., & Gramling, R. (1993). Socioenvironmental factors and development policy: Understanding opposition and support for offshore oil. *Sociological Forum, 8*(3), 341–364. doi:10.1007/BF01115049
Freudenburg, W. R., & Gramling, R. (1994). *Oil in troubled waters: Perceptions, politics, and the battle over offshore drilling.* Albany: State University of New York Press.

Freudenburg, W. R., & Jones, T. R. (1991). Attitudes and stress in the presence of technological risk: A test of the Supreme Court hypothesis. *Social Forces, 64*(4), 1143–1168.
Freudenburg, W. R., & Pastor, S. (1992). Public responses to technological risks: Toward a sociological perspective. *The Sociological Quarterly, 33*(3), 389–412.
Furedi, F. (2009). Precautionary culture and the rise of probabilistic risk assessment. *Erasmus Law Review, 02*(02), 197–220.
Giddens, A. (1990). *The consequences of modernity.* Stanford, CA: Stanford University Press.
Giddens, A. (1991). *Modernity and self-identity: Self and society in the late modern age.* Stanford, CA: Stanford University Press.
Giddens, A. (1994). *Beyond left and right: The future of radical politics.* Cambridge: Polity Press
Goss, C., & Kamieniecki, S. (1983). Congruence between public opinion and congressional actions on energy issues, 1973–1974. *Energy Syst. Policy, 7*(2), 149–170.
Gross, P. R., & Levitt, N. (1994). *Higher superstition: The academic left and its quarrels with science.* Baltimore, MD: Johns Hopkins University Press.
Habermas, J. (1970). *Toward a rational society: Student protest, science, and politics.* Boston: Beacon Press.
Habermas, J. (1973). *Legitimation crisis.* Boston: Beacon Press.
Hansen, S. F., Maynard, A., & Baun, A. (2008). Late lessons from early warnings for nanotechnology. *Nature Nanotechnology, 3,* 444–447.
Heberlein, T. (1981). Environmental attitudes. *Zeitschrift fur Umweltpolitik 2/81,* 241–270.
Kahan, D. M., & Rejeski, D. (2009). Toward a comprehensive strategy for nanotechnology risk communication. *Project on Emerging Nanotechnologies: Research Brief, Pen Brief No. 5.* Retrieved December 15, 2011, from: http://www.culturalcognition.net/storage/nano_090225_research_brief_kahan_nl1.pdf
Kahan, D. M., Slovic, P., Braman, D., Gastil, J., & Cohen, G. L. (2007, March 7). *Affect, values, and nanotechnology risk perceptions: an experimental investigation* (GWU Legal Studies Research Paper No. 261; Yale Law School, Public Law Working Paper No. 155; GWU Law School Public Law Research Paper No. 261; 2nd Annual Conference on Empirical Legal Studies Paper). Retrieved December 2, 2011, from http://ssrn.com/abstract=968652
Kroll-Smith, J. S., & Couch, S. R. (1993). Symbols, ecology, and contamination: Case studies in the ecological-symbolic approach to disaster. R*esearch in Social Problems and Public Policy, 5,* 47–73.
La Porte, T. R., & Metlay, D. (1975). Technology observed: Attitudes of a wary public. *Science, 188*(4184), 121–127.
Lang, J. T., & Hallman, W. K. (2005). Who does the public trust? The case of genetically modified food in the United States. *Risk Analysis, 25*(5), 1241–1252.
List of nanotechnology applications. (n.d.). In *Wikipedia.* Retrieved August 30, 2011, from http://en.wikipedia.org/wiki/List_of_nanotechnology_applications.
Lopes, L. L. (1991). The rhetoric of irrationality. *Theory & Psychology, 1*(1), 65–82. doi:10.1177/0959354391011005
Macoubrie, J. (2006). Nanotechnology: Public concerns, reasoning and trust in government. *Public Understanding of Science, 15,* 221–224.
Marchant, G. E., & Sylvester, D. J. (2006). Transnational models for regulation of nanotechnology. *The Journal of Law, 2006*(Winter), 2–13.
Marchant, G. E., Sylvester, D. J., & Abbott, Kenneth W. (2009, August 4). What does the history of technology regulation teach us about nano oversight? *The*

Journal of Law, Medicine & Ethics. Retrieved December 2, 2011, from http://ssrn.com/abstract=1470446

Mazur, A. (1981). Media coverage and public opinion on scientific controversies. *Journal of Communication, 31*(2), 106–115.

Merton, R. K. (1936). The unanticipated consequences of purposive social action. *American Sociological Review, 1*(6), 894–904.

O'Connor, J. (1973). *The fiscal crisis of the state*. New York: St. Martin Press.

Offe, C. (1987). *Contradictions of the welfare state*. Cambridge, MA: MIT Press.

Perrow, C. (1984). *Normal accidents: Living with high-risk technologies*. New York: Basic Books.

Peter D. Hart Research Associates. (2009). *Nanotechnology, synthetic biology, and public opinion*. Workshop summary.

Petersen, A., Anderson, A., Wilkinson, C., & Allan, S. (2007). Nanotechnologies, risk and society. *Health, Risk & Society, 9*(2), 117–124. doi:10.1080/13698570701306765

Picou, J. S., & Gill, D. A. (1999). The Exxon Valdez disaster as localized environmental catastrophe: Dissimilarities to risk society theory. In M. J. Cohen (Ed.), *Risk in the modern age: Social theory, science and environmental decision-making*, pp. 143–170. New York: St. Martin's Press.

Picou, J. S., Gill, D. A., Dyer, C. L., & Curry, E. W. (1992). Disruption and stress in an Alaskan fishing community: Initial and continuing impacts of the Exxon Valdez oil spill. *Organization & Environment, 6*, 235–257.

Picou, J. S., Marshall, B. K., & Gill, D. A. (2004). Disaster, litigation, and the corrosive community. *Social Forces, 82*(4), 1493–1522.

Pidgeon, N., Harthorn, B. H., Bryant, K., & Rogers-Hayden, T. (2009). Deliberating the risks of nanotechnologies for energy and health applications in the United States and United Kingdom. *Nature Nanotechnology, 4*, 95–98.

Piegorsch, W. W., & Schuler, E. (2008). Communicating the risks, and the benefits, of nanotechnology. *International Journal of Risk Assessment and Management, 10*(1–2), 57–69.

Pijawka, K. D., & Mushkatel, A. H. (1991). Public opposition to the siting of the high-level nuclear waste repository: The importance of trust. *Policy Studies Review, 10*(4), 180–194.

Prades, A., Espluga, J., Real, M., & Sola, R. (2009). The siting of a research centre on clean coal combustion and CO_2 capture in Spain: Some notes on the relationship between trust and lack of public information. *Journal of Risk Research, 12*(5), 709–723.

Reed, J. H., & Wilkes, J. M. (1981). Technical nuclear knowledge and attitudes toward nuclear power before and after Three Mile Island. Paper presented at the Annual Meeting of the Society for the Social Study of Science, Atlanta, GA.

Rogers-Hayden, T., & Pidgeon, N. (2008). Developments in nanotechnology public engagement in the UK: "Upstream" towards sustainability. *Journal of Cleaner Production, 16*, 1010–1013.

Royal Society & Royal Academy of Engineering. (2004, July). *Nanoscience and nanotechnologies: Opportunities and uncertainties*. Cardiff, UK: Clyvedon Press.

Satterfield, T., Kandlikar, M., Beaudrie, C. E. H., Conti, J., & Harthorn, B. H. (2009). Anticipating the perceived risk of nanotechnologies. *Nature Nanotechnology, 4*, 752–758.

Schuler, E. (2004). Perception of risks and nanotechnology. In D. Baird, A. Nordmann, & J. Schummer (Eds.), *Discovering the nanoscale*, pp. 279–284. Amsterdam: IOS Press.

Short, J. F. (1984). The social fabric at risk: Toward the social transformation of risk analysis. *American Sociological Review, 49*(6), 711–725.

Siegrist, M. (2000). The influence of trust and perceptions of risks and benefits on the acceptance of gene technology. *Risk Analysis, 20*(2), 195–203.

Siegrist, M., Earle, T. C., & Gutscher, H. (2003). Test of a trust and confidence model in the applied context of electromagnetic field (EMF) risks. *Risk Analysis, 23*(4), 705–716.

Siegrist, M., Keller, C., Kastenholz, H., & Frey, S. (2007). Laypeople's and experts' perception of nanotechnology hazards. *Risk Analysis, 27*(1), 59–69.

Slovic, P. (1987). Perception of risk. *Science, 236,* 280–285.

Slovic, P., Layman, M., & Flynn, J. H. (1993). Perceived risk, trust, and nuclear waste: Lessons from Yucca Mountain. In R. E. Dunlap, M. E. Kraft, & E. A. Rosa (Eds.), *Public reactions to nuclear waste: Citizens' views of repository siting*, pp. 64–86. Durham, NC: Duke University Press.

Stone, C. N. (1980). Systemic power in community decision making: A restatement of stratification theory. *The American Political Science Review, 74*(4), 978–990.

Suppan, S. (2011). *Racing ahead: US agri-nanotechnology in the absence of regulation.* Minneapolis, MN: Institute for Agriculture and Trade Policy.

Sylvester, D. J., Abbott, Kenneth W., & Marchant, G. E. (2009). Not again! Public perception, regulation, and nanotechnology. *Regulation & Governance, 3,* 165–185.

Waldron, A. M., Spencer, D., & Batt, C. A. (2006). The current state of public understanding of nanotechnology. *Journal of Nanoparticle Research, 8,* 569–575.

Weber, M. (1946). Science as a vocation. In H. Gerth & C. W. Mills, (Eds.) *Essays in sociology*, pp. 129–156. London: Routledge & Kegan Paul. (Original work 1919)

Wilkinson, C., Allan, S., Anderson, A., & Petersen, A. (2007). From uncertainty to risk? Scientific and news media portrayals of nanoparticle safety. *Health, Risk & Society, 9*(2), 145–157. doi:10.1080/13698570701306823.

Contributors

EDITORS

BARBARA HERR HARTHORN

Barbara Herr Harthorn is a cultural anthropologist who is associate professor of feminist studies, anthropology, and sociology; and founding director and, since 2005, lead principal investigator (PI) of the National Science Foundation–funded Nanoscale Science and Engineering Center: Center for Nanotechnology in Society at University of California, Santa Barbara. In CNS-UCSB she leads an interdisciplinary group studying nanotechnology risk perception, public deliberation, and issues of inequality. She is also co-PI on the UCSB award in the NSF/EPA-funded UC Center for Environmental Implications of Nanotechnology (UCLA) where she leads a social science group studying environmental risk perception. She completed her BA in Anthropology from Bryn Mawr College, and her MA and PhD in medical and psychological anthropology at UCLA before pursuing postdoctoral training in social psychology at UCSB. She is author (with Laury Oaks) of *Risk, Culture, and Health Inequality: Shifting Perceptions of Danger and Blame* (Greenwood/Praeger, 2003). She is a fellow of the American Association for the Advancement of Science.

JOHN W. MOHR

John W. Mohr is professor of sociology at the University of California, Santa Barbara. He received his PhD in sociology at Yale University. He has long been interested in understanding the cultural logic of institutional fields. Earlier studies focused on the history of the American welfare state and recent battles over the discourse of affirmative action in California. Along with Roger Friedland he is the organizer of the Cultural Turn Conference series at UCSB and the co-editor of *Matters of Culture* (Cambridge University Press, 2004). Mohr has published a number of articles on methods of institutional analysis, including "Measuring Meaning Structures" (*Annual Review of Sociology*), "Four Ways to Measure Culture" (with Craig Rawlings) (in *The Oxford Handbook of Cultural Sociology*), and "The Duality of Culture and Practice" (with

Vincent Duquenne) and "How to Model an Institution" (with Harrison White) (both published in *Theory and Society*). His current research projects include a study of faculty activism in higher education and an analysis of the discursive logic of national security institutions.

CONTRIBUTORS

RICHARD P. APPELBAUM

Richard P. Appelbaum is professor and MacArthur chair in global and international studies and sociology at the University of California, Santa Barbara. He is co-PI at the National Science Foundation–funded Nanoscale Science and Engineering Center, Center for Nanotechnology in Society, where he directs the interdisciplinary research group on globalization and nanotechnology.

BRUCE BIMBER

Bruce Bimber is professor of political science at the University of California, Santa Barbara, where he is also founder and former director of the Center for Information Technology and Society. He studies political communication, with an emphasis on relationships between digital media use, collective action, and political behavior. His interest in digital media and society arises from his training as an electrical engineer, and from many years of observing the interconnections between social and technological innovation. His books include *Collective Action in Organizations* (with Andrew Flanagin and Cynthia Stohl) (Cambridge University Press, 2012); *Campaigning Online: The Internet in US Elections* (with Richard Davis) (Oxford University Press, 2003), and *Information and American Democracy* (Cambridge University Press, 2003). Bimber is a former fellow of the Center for Advanced Study in the Behavioral Sciences, and a fellow of the American Association for the Advancement of Science.

JOHN SEELY BROWN

John Seely Brown was the Chief Scientist of Xerox Corporation and the director of its Palo Alto Research Center (PARC) for nearly two decades. He is an author of some 200 articles, papers and chapters and co-author of 7 books including *The Power of Pull: How Small Moves, Smartly Made, Can Set Big Things in Motion*, Basic Books, 2010 (with John Hagel and Lang Davison) and *The Social Life of Information*, Harvard Business School, 2000 (with Paul Duguid). He received his PhD from

University of Michigan in computer and communication sciences and is currently a visiting scholar and advisor to the Provost at the University of Southern California (USC).

CONG CAO

Dr. Cong Cao (PhD in Sociology, Columbia University) is associate professor and reader of contemporary Chinese studies at the University of Nottingham. Educated in both China and the United States, and in both natural and social sciences, he has worked at the University of Oregon, the National University of Singapore, and the State University of New York. As one of the leading scholars in the studies of China's science, technology, and innovation, Dr. Cao is the author of *China's Scientific Elite* (RoutledgeCurzon, 2004), a study of those Chinese scientists holding the elite membership in the Chinese Academy of Sciences, and *China's Emerging Technological Edge: Assessing the Role of High-End Talent* (with Denis Fred Simon) (Cambridge University Press, 2009). His papers have appeared in the leading international journals of social studies of science and technology, and of China and Asian studies.

MARY B. COLLINS

Mary is a PhD candidate at the Bren School of Environmental Science and Management at the University of California, Santa Barbara. Broadly, Mary studies environment–society systems with a twofold focus: environmental risk perception and environmental inequality. Mary's risk perception research is in conjunction with the UC Center for the Environmental Implications of Nanotechnology; it evaluates public risk perception in the context of emergent technology. Mary's environmental inequality research involves combining large social and environmental datasets in an effort to understand both the ecological and political/economic forces that drive contamination events in human-dominated ecosystems. Mary received her bachelor's degree in sociology from the University of Wisconsin in Madison and her master's degree in applied sociology at the University of Central Florida in Orlando.

MEREDITH CONROY

Meredith's research focuses on the communication of issues in news media, and its effect on public attitudes. As a political psychologist, Meredith is especially interested in understanding the impact of informational frames and comparisons disseminated by the media on the stability and strength of individual attitudes toward issues. Meredith has a PhD from University of California, Santa Barbara, a master's degree from Purdue University, and a bachelor's degree from Whittier College, all in political science. She is currently teaching at Occidental College, in Los Angeles, CA.

ADAM CORNER

Adam Corner is a research associate in the Understanding Risk research group at Cardiff University. His research looks at the communication of climate change, how people evaluate arguments and evidence, and public engagement with emerging areas of science such as nanotechnologies and geoengineering.

MATTHEW N. EISLER

Matthew N. Eisler's general research interests are in the history of the social relations of science and engineering and their material effects on natural and built spaces, especially after 1945. He has explored the relationship between linear ideology/project management, utopian/futurist discourse, patronage, and the construction of expert authority in case studies of fuel cell research and development and the U.S. Department of Energy's Nanoscale Science, Engineering, and Technology (NSET) programs. Eisler obtained a PhD in history from the University of Alberta, was the Harris Steel Postdoctoral Fellow at the Department of History at the University of Western Ontario (2008–2009), and a postdoctoral fellow at the Center for Nanotechnology in Society at the University of California, Santa Barbara from 2009 to 2011. He is currently research fellow at the Center for Contemporary History and Policy at the Chemical Heritage Foundation in Philadelphia.

WILLIAM R. FREUDENBURG

William Freudenburg (1951–2010) was an environmental sociologist and social theorist who received his Ph.D. from Yale University (1979) and taught most recently at the University of California, Santa Barbara where he was the Dehlsen Professor of Environmental Studies. He published some 200 research articles, chapters and technical reports and co-authored or co-edited 10 books including, *Catastrophe in the Making: The Engineering of Katrina and the Disasters of Tomorrow*, Island Press, 2009 (with Robert Gramling, Shirley Laska and Kai Erikson) and, most recently, *Blowout in the Gulf: The BP Oil Spill Disaster and the Future of Energy in America, MIT*, 2010 (with Robert Gramling).

MIKAEL JOHANSSON

Mikael Johansson got his PhD at the Department of Social Anthropology at Goteborg University, Sweden. His dissertation on the cosmology of nanoscientists utilized lab ethnography as well as historical context to examine the "scientist's view" of nanotechnology. During 2009–2010 Mikael was a CNS-UCSB postdoctoral scholar in which he did a ethnographic fieldwork to further the field of cross-cultural analysis of nanoscientific research groups, including focus on concepts such as risk, laboratory space, and globalization. Mikael has now returned to the Department of Global Studies, Gothenburg University, and lectures part-time and is continuing his anthropological research on nanoscientists.

ERICA LIVELY

Erica Lively is pursuing a PhD in electrical engineering at the University of California, Santa Barbara. She obtained her BS in electrical engineering from the University of Idaho in 2005 and MS in electrical engineering from the UC Santa Barbara in 2007. Erica was a graduate fellow from 2007 to 2010 at the Center for Nanotechnology in Society at UCSB. She was also a Mirzayan Science and Technology Policy Fellow at the National Academy of Engineering in 2011. Erica's research interests include the nanophotonics, as well as the political and societal implications of science and technology.

TYRONNE MARTIN

Tyronne Martin is a doctoral student in the Bren School of Environmental Science and Management and a master's student in the Department of Chemistry and Biochemistry at the University of California, Santa Barbara. He is a former science and engineering graduate fellow at the UCSB Center for Nanotechnology in Society. As a fellow he contributed by informing the public and observing their perceptions of benefits and risks regarding nanotechnology. He is currently a researcher for the Center for Environmental Implications of Nanotechnology (CEIN) at UCSB and his research areas include: cellular toxicology, mechanisms of toxicity, and the ecological impacts of nanomaterials on marine systems.

W. PATRICK MCCRAY

W. Patrick McCray is a professor in the Department of History at the University of California, Santa Barbara. In 2011–2012, he was also the Eleanor Searle Visiting Professor in the History of Science at the California Institute of Technology. McCray entered the historians' profession via his original career as a scientist. He has written widely on the history of science and technology after 1945. His book *Giant Telescopes: Astronomical Ambition and the Promise of Technology* (Harvard University Press, 2004) explored how scientists build and use today's most modern telescopes. A subsequent project examined the activities of citizen-scientists during the Cold War: *Keep Watching the Skies: The Story of Operation Moonwatch and the Dawn of the Space Age* (Princeton University Press, 2008). His new book, tentatively titled *Limitless: From Space Colonies to Nanotechnologies in Pursuit of the Future*, is about people who used their expertise as scientists, engineers, and popularizers to promote visions of a more expansive technological future (Princeton University Press, 2012).

CYRUS C. M. MODY

Cyrus C. M. Mody is an assistant professor in the History Department at Rice University. He is the author of *Instrumental Community: Probe Microscopy and the Path to Nanotechnology* (MIT Press, 2011) as well as numerous articles on the histories of subfields of nanotechnology including molecular electronics, fullerene chemistry, and microfabrication.

YASUYUKI MOTOYAMA

Yasuyuki Motoyama was a postdoctoral scholar at the Center for Nanotechnology in Society at University of California, Santa Barbara between 2008 and 2011. He received his PhD in city and regional planning at University of California, Berkeley, and his dissertation investigated the engineering and organizational factors that affected geographic concentration of product development by multinational firms. At UC Santa Barbara, his projects included political economy of science and industrial policy in the United States, geographic analysis of Chinese nanotechnology patents, commercialization of university technology in Japan, and bibliometric analysis of nanotechnology articles and patents. Since 2011, he has been a senior scholar at the Ewing Marion Kauffman Foundation and conducts research economic development through entrepreneurship, innovation, and technology commercialization.

CHRISTOPHER NEWFIELD

Christopher Newfield teaches American studies in the English Department at the University of California, Santa Barbara. His current research focuses on higher education history, funding, and policy, culture and innovation, and the relation between culture and economics. Recent articles have appeared in the *Chronicle of Higher Education*, *Academe*, Le *Monde Diplomatique*, *La Revue Internationale des Livres et des Ideés*, *Radikal* (Turkey), *Social Text*, *Critical Inquiry*, and *South Atlantic Quarterly*, and include "The View from 2020: How Universities Came Back," "The End of the American Funding Model: What Comes Next?" "Ending the Budget Wars: Funding the Humanities during a Crisis in Higher Education," "Public Universities at Risk: 7 Damaging Myths," "Science and Social Welfare," "Why Public is Losing to Private in American Research," and "Can American Studies Do Economics?" He is the author of *The Emerson Effect: Individualism and Submission in America* (University of Chicago Press, 1996), *Ivy and Industry: Business and the Making of the American University, 1880–1980* (Duke University Press, 2003), and *Unmaking the Public University: The Forty Year Assault on the Middle Class* (Harvard University Press, 2008). He chairs the Innovation Group at the NSF Center for Nanotechnology in Society, runs a blog on the current crisis in higher education, *Rethinking the University* (http://utotherescue.blogspot.com), blogs at the Huffington Post, and is working on a book called *Lower Education: What to Do About Our Downsized Future.*

RACHEL PARKER

Rachel Parker's recent research at the Science & Technology Policy Institute (STPI) has focused on the intersection of science and technology with research capacity development. Her work at STPI includes research capacity development-focused efforts for the National Institutes of Health and the Office of Science and Technology Policy as well as the National Science Foundation. Rachel works on a variety of evaluation-focused projects at STPI and continues to write about nano and other emerging technologies and economic development in China and, more recently, in Mexico. With Richard Appelbaum, Rachel is editor of a new volume, *Can Emerging Technologies Make a Difference in Development?* (Routledge, 2012). Rachel received a bachelor's degree in sociology from Brandeis University, an MSc in management of non-governmental organizations from the London School of Economics and Political Science, and a PhD in sociology with an emphasis in global and international studies from the University of California, Santa Barbara.

NICK PIDGEON

Nick Pidgeon is professor of environmental psychology at Cardiff University, where he directs the Understanding Risk Research Group (see www.understanding-risk.org). His research looks at the public acceptability and societal governance of environmental and technological risks, including nuclear power, climate change, and the emerging technologies of nanotechnology and geoengineering. He was a member of the Royal Society & Royal Academy of Engineering nanotechnology study group and is currently a member of the Science Advisory Group of the UK's Department for Energy and Climate Change. Coeditor with Roger Kasperson and Paul Slovic of *The Social Amplification of Risk* (Cambridge University Press, 2003) and with the late Barry Turner of the 2nd edition of *Man Made Disasters* (Butterworth Heinemann, 1997).

JENIFER ROGERS-BROWN

Jennifer Rogers-Brown is an assistant professor of sociology at Long Island University, CW Post. She received her PhD in Sociology from the University of California, Santa Barbara. Her dissertation analyzes maize in Mexico, free trade, bioengineering, gender, and indigenous cultures. She has also been a visiting researcher and postdoctoral scholar at the Center for Nanotechnology in Society at UC Santa Barbara where she studied public perceptions of nanotechnology and gender. She also studies gender, technology, and resistance to agricultural technologies in the United States and Mexico.

CHRISTINE SHEARER

Christine Shearer is a postdoctoral scholar at the Center for Nanotechnology in Society at UC Santa Barbara. Her research focuses on environmental sociology and science, technology, and society, and she is author of the book, *Kivalina: A Climate Change Story* (2011). She holds a PhD in Sociology from UC Santa Barbara.

DAVID WEAVER

David A. Weaver holds a doctorate in political science from UC Santa Barbara, where he specialized in American politics and political communication. He was a Graduate Fellow at the Center for Nanotechnology in Society (CNS-UCSB) from 2006 to 2008 where he collaborated on research related to media framing of nanotechnologies. He also holds a MA in mass communication from the University of North Carolina at Chapel Hill. His current research focuses on media framing of political activism as well as analyzing audiences of political talk radio. He has previously published in *Journalism and Mass Communication Quarterly* and *Science Communication*. He is currently instructor of political science and communication at Boise State University.

Index

O

P

Q

R